BRITISH RAIL FLEET S

3 Production Diesel-Electrics Types 4 and 5

Brian Haresnape revised by S. R. Batty

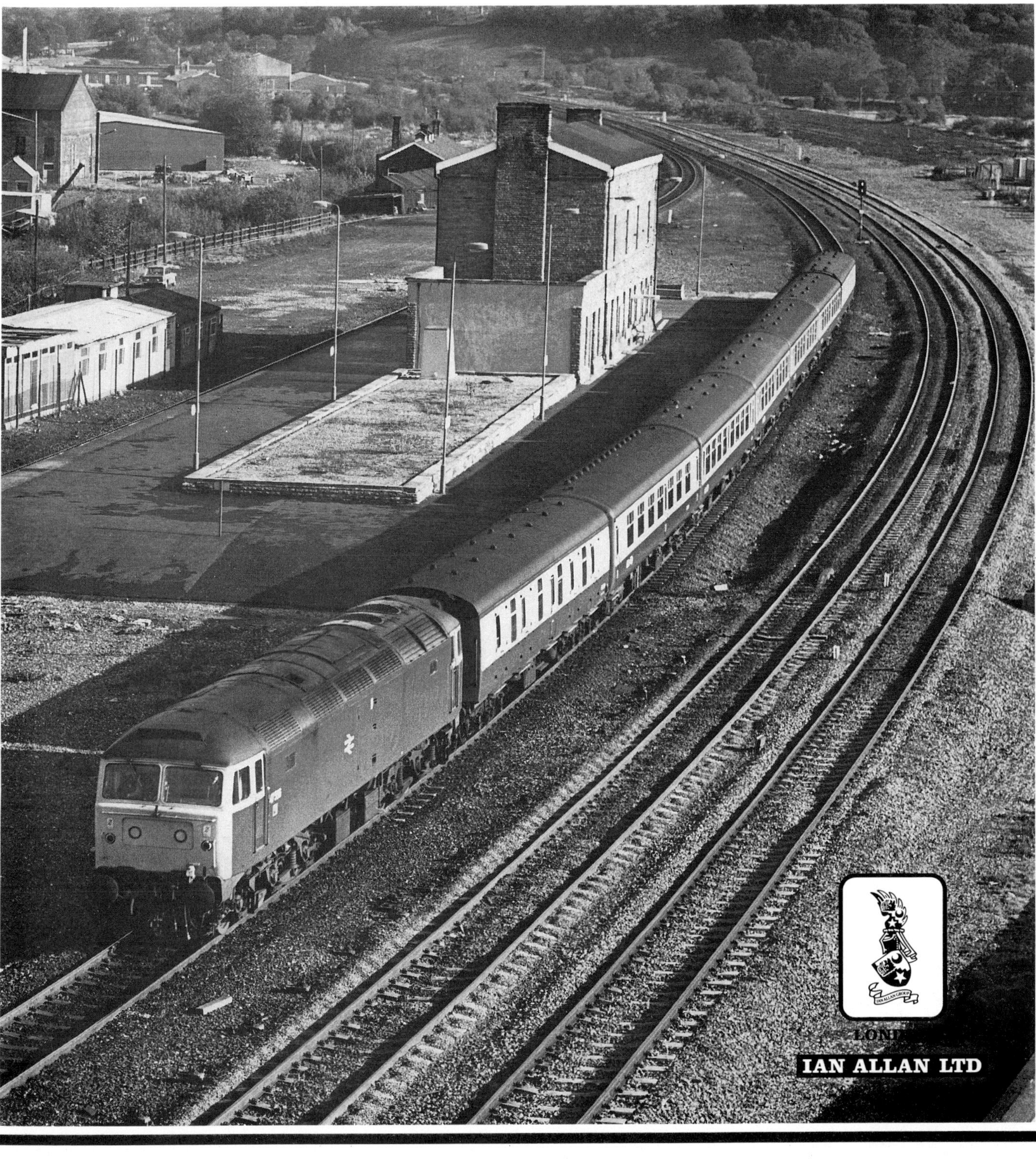

LOND
IAN ALLAN LTD

First published 1982
This edition 1989

ISBN 0 7110 1859 6

Published by Ian Allan Ltd, Shepperton, Surrey; and printed by Ian Allan Printing Ltd at their works at Coombelands in Runnymede, England

Previous page:
Class 47 No 47536 passes Mirfield with the 11.00 Scarborough-Bangor on 24 October 1983. *S. R. Batty*

Introduction to the Third Edition

Several years have passed since this volume of the *Fleet Survey* series was last published, and this revised third edition brings the story of BR's Type 4 and 5 classes up to 1989. Many changes have taken place — several classes are now extinct — and I hope the completely revised pictorial content of the album will give the reader an interesting and informative glimpse of the 'top end' of the locomotive fleet.

In the natural order of things, Brian Haresnape would, of course, have updated his own original work to produce this edition, but his tragic and violent death has robbed the railway enthusiast of a talented artist who was also an excellent raconteur of the BR diesel traction scene. I have left much of Brian's text unaltered, making changes only for the sake of updating events, and I am pleased to have been able to play a small part in furthering the memory of a well-respected railway author. My thanks to all concerned.

S. R. Batty

Front cover:
The Network SouthEast livery carried by Class 50 No 50003 *Temeraire* makes a rather incongruous sight at Bodmin Parkway on 6 August 1988. The locomotive is working the 08.17 Penzance-Glasgow InterCity service.
Peter J. Robinson

Back cover:
Class 58 heavy freight locomotive No 58050 in the paint shop at Doncaster Works on 26 January 1987. Class 58s are now receiving Railfreight Sector grey livery. *Les Nixon*

Below:
Class 47/0 No 47231 *The Silcock Express*, named after the Merseyside-based automotive distribution company, makes a fine sight near Manea with the 10.08 Whitemoor-Norwich Railfreight Distribution service on 22 June 1988. The locomotive is in the latest Railfreight livery. *John C. Baker*

Contents

The Quest for Efficiency

In the first two books of this series I have described the diesel-electric locomotives produced for the British Transport Commission pilot scheme orders, and their diesel-hydraulic equivalents (and later production orders) on the Western Region. This book continues to the present day the story of the largest diesel-electric locomotives, designated as Types 4 and 5. These locomotives can conveniently be described as the 'production series'. In two instances they were simply the improved production version of the original pilot scheme designs, in others they were completely new designs, produced as the direct result of the harsh experience gained by British Railways in the early days of dieselisation. They represent the powerful mainstay of the contemporary British Rail fleet.

I have already described in *1: Early Prototype and Pilot Scheme Diesel-Electrics* how events overtook the original pilot scheme plan, and how it became imperative for the BTC to place large repeat orders for a wide variety of designs, in some cases before even a single example had been completed and tested in service. In the case of the largest pilot scheme types, originally Type C and later Type 4, the range was fortunately limited, and in some details the two diesel-electric designs — the EE Co D200 class and the BR/Sulzer D1 class 1Co-Co1s, of 2,000 and 2,300hp respectively — were basically similar. Only the WR B-B 'Warship' class 2,200hp diesel-hydraulic, produced by Swindon to German specification represented an unorthodox solution to the more powerful range of traction required for express passenger and fast freight duties, being of lightweight stressed-steel construction with two quick-running engines (see *2: The Western Region Diesel-Hydraulics*).

The two diesel-electric designs were of very robust conception indeed (as has been well-proven by their longevity in BR service). As we have seen in the first book of this series, their design arose directly from the experience gained with the pioneer LMSR Ivatt Co-Cos and their SR Bulleid 1Co-Co1 counterparts.

Briefly to recap the situation leading to their introduction, it was the completion of the third SR locomotive, at Brighton works in March 1954, that heralded both the end of a brief diesel development era (1947-54) and also paved the way for these future designs.

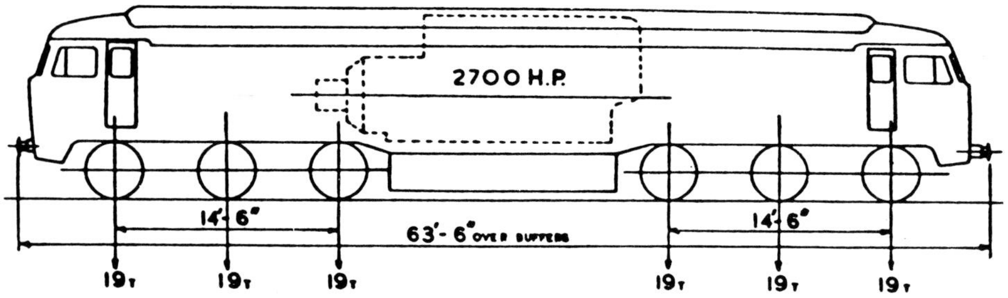

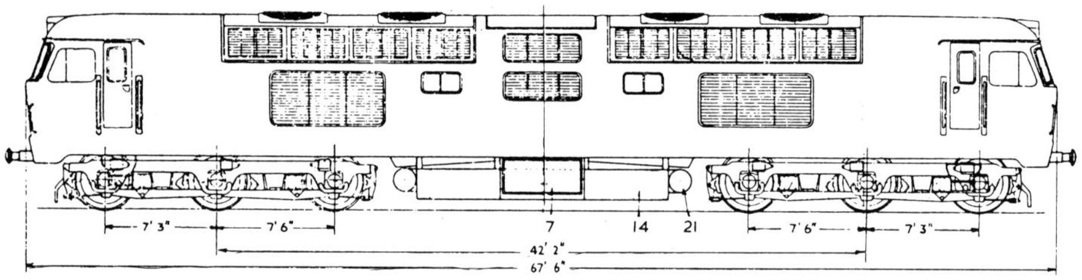

Top:
The BR outline specification for a second generation Type 4 design, with six axles loaded to 19ton apiece and a 2,700hp diesel engine. The outline of the cabs and roof followed Design Panel recommendation on appearance.

Above:
The proposed 4,400hp 'Super-Deltic' design, with BR-inspired front end. This would have been a most impressive machine.

Below:
Outline of the Class 58 heavy duty freight locomotive which superseded the Class 56 in production. The completely altered body layout, with what is in effect a hood unit sandwiched between two full-width cabs, closely resembles contemporary American thinking for such locomotives, and was evolved partially with an eye to the export market for British Rail Engineering Ltd.

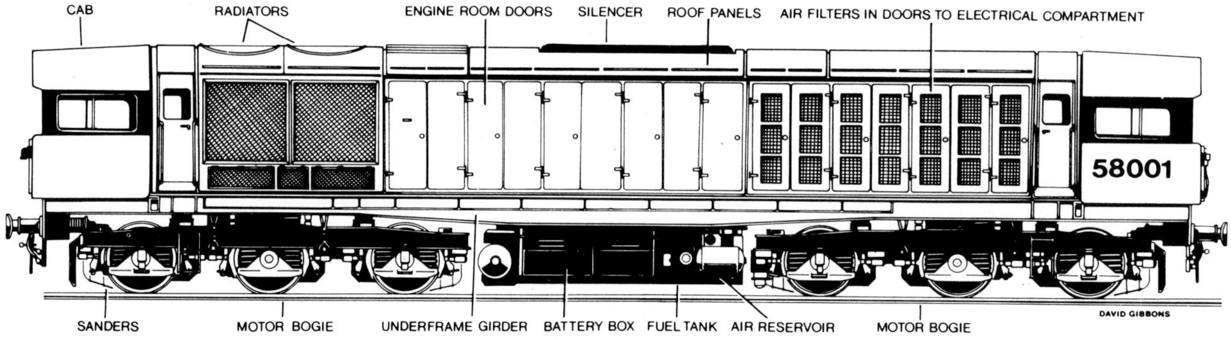

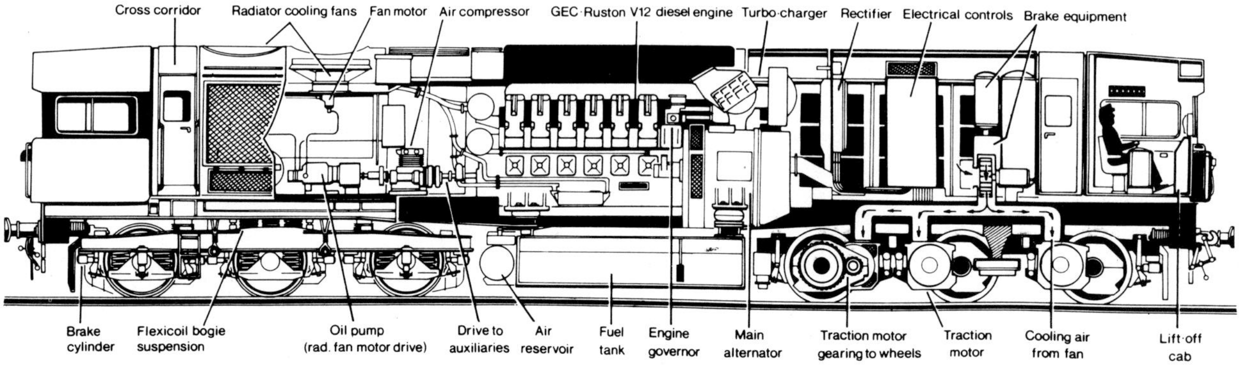

Top:
The Sunday 17.30 St Pancras-Manchester Piccadilly passes Castle Donington on 22 July 1962 behind an unidentified English Electric Type 4 locomotive. The venerable 16SVT engine was coaxed up to 2,000hp for use in this 133-ton machine, and gave reliable service for a quarter of a century. *M. Mitchell*

Above:
Access between coupled locomotives was soon realised largely to be unnecessary, and later examples of the EE Co product were fitted with a much neater central headcode panel. No D358 passes Craigentinny with the 11.45 Edinburgh Waverley-Newcastle on 22 August 1962. *A. A. Vickers*

Left:
The Class 40s suffered undue neglect during their later lives, probably due to a lack of available money for repairs. No 40069, with bent cab steps, speeds along between Sessay and Thirsk with a down ECML express on 15 July 1978. This locomotive was unusual in having had a section of bodyside valancing removed during the 1960s to allow the oil drainage pipes to be experimentally repositioned. *J. E. Oxley*

Below:

The BR/Sulzer Type 4 design utilised an uprated 2,500hp version of the LDA28 engine used in the D1-D10 series, but the 'Peaks' still turned the scales at 138 tons. Nos D16, D12 and Type 2 No D5149 are seen at Leeds Neville Hill depot on 8 June 1961, with D12 showing the original front-end design and D16 having the gangway doors omitted but still using twin headcode boxes. Note also the complete absence of any yellow warning paint! *G. W. Morrison*

A

Above:
Later 'Peaks' were fitted with a simple one-piece central panel, as seen on No 45108 accelerating away from Newcastle with empty stock for Heaton sidings on 7 August 1973. *T. Koerner*

For a while steam once again reigned supreme on BR, with many new Standard locomotives constructed each year. However, as early as December 1954 the BTC published its plan for 'Modernisation and Re-equipment of British Railways', in booklet form. This said, amongst other things that '. . . major changes in the forms of motive power on British Railways must be effected, involving a widespread changeover from steam to electricity and diesel power'. They favoured electrification, with just reasoning, but diesel traction was seen to offer many of the advantages of electricity in terms of cleanliness, acceleration and economical operation. Diesel traction was seen as the ideal solution for lines where traffic was not sufficiently dense to warrant the high initial cost of electrification, and also as a stopgap solution on some main lines pending electrification.

An area-by-area conversion from steam to diesel was proposed as an immediate need. Use of diesel railcars was already being implemented and the railcar fleet was steadily growing. The plan envisaged a fleet of some 2,500 main line diesel locomotives on BR by 1970, of highly standardised design and mostly of 2,000hp. At the time the plan was published there were approximately 19,000 steam locomotives in service and by eliminating steam area by area the BTC wisely hoped for a minimum of complication in the operating and maintenance areas of diesel traction — a dream that, alas, was not to come true for a good many years!

The plan envisaged all steam abolished west of Newton Abbot on the WR, and that the Waterloo-Exeter, and the Waterloo-Southampton-Bournemouth-Weymouth routes should also be early candidates. The WR

Right:
The Trans-Pennine (Northern) route was a 'Peak' stronghold for many years, and the class played out their main line service on these duties. An unidentified Class 45/1 approaches Heaton Lodge with the 10.03 Liverpool-Scarborough on 8 September 1986. *S. R. Batty*

Above:
Of all the prototypes tested by BR, the BRCW/Sulzer *Lion* was certainly the most startling in appearance, with all-white livery and fluted bodysides. With its single 2,750hp engine and BR-preferred cab design, this machine was almost put into series production as the new standard Type 4 design, but for unknown reasons the proposal was dropped in favour of the Brush/Sulzer product. *AEI*

scheme came to fruition but with diesel-hydraulic traction, which was not greatly favoured by BTC central engineering chiefs.

On a visit to Germany the BTC and WR chief mechanical engineers had been sufficiently impressed by diesel-hydraulic traction to agree upon a small experiment with some locomotives of this type on the WR, but the future was clearly seen as predominantly one for diesel-electric transmission. How the

Below:
The Brush *Falcon* design also employed the now-familiar cab design, but was powered by two Bristol-Siddeley/Maybach high speed engines producing a total of 2,800hp. Despite having successful trials, BR at the time did not favour fast-running engines and the design was not progressed. *Falcon* is seen here, in its original livery of pale green and chestnut brown, at Finsbury Park on 13 October 1961. *BR*

Above:
***Falcon's* usefulness ensured its survival until the mid-1970s, and it is seen here in BR blue with full yellow ends, at Ebbw Junction on 24 April 1975. Numbered 1200 (the only member of Class 53), this was originally to have been the basis for a small fleet of Crewe-built Co-Co type locomotives.**
P. D. Hawkins

WR then succeeded in an initial conversion to diesel-hydraulics, only to have their entire fleet replaced by diesel-electrics, is the saga told in the second book of this series.

The initial pilot scheme orders were agreed upon to allow various types of diesel engine and transmission equipment to be evaluated, and to give a number of locomotive builders an opportunity to show their abilities in the changeover from steam to diesel. From the experience gained from the use of these 174 varied locomotives it was intended that a limited range of standard designs should be drawn-up and constructed for the area-by-area conversion. Meanwhile BR had only the five existing main line diesels to give them some limited operational experience, plus the very experimental Derby/Fell mechanical drive 4-8-4 No 10100 (to be described in detail in a later volume) and these five main line diesels all had electric transmission. Despite far from ideal operating conditions, their design was to prove sufficiently satisfactory to allow the BTC to think ahead in terms of 2,000-2,300hp 1Co-Co1 diesel-electrics, largely based upon their design, as the future heavy duty main line locomotive, suitable for both express passenger and freight use. These two pilot scheme designs had, as already stated, many features in common, in particular the use of large slow running diesel engines and cumbersome 1Co-Co1 bogies. They were heavyweight machines, partly because it was considered necessary to provide them with sufficient brake force to permit the working of partially brake-fitted freights, and partly because basically traditional methods of construction were employed.

As described in Book One, the English Electric Co was determined to produce a main line diesel-electric locomotive of greatly improved power/weight ratio, and this emerged as a private venture in 1955 in the form of the 3,300hp prototype 'Deltic' with two quick-running Napier engines. Trial running commenced on the LMR, but no further Co-Co 'Deltics' featured in the pilot scheme orders. Instead, for some odd reason, the Napier engine was selected for use (one per locomotive) in 10 Type 2 Bo-Bos; an application which could not reap the maximum benefits it had to offer!

So we now take up the story from the time when financial pressures forced the BTC to make the decision to abandon the three-year pilot scheme programme and to accelerate the introduction of diesel power, even before the first of the new locomotives had been completed (in late 1956). The BTC Chief Mechanical and Electrical Engineers, R. C. Bond and S. B. Warder, were faced with a most difficult situation. The BTC, understandably, wanted to limit the number of types selected for further quantity production, in an attempt to retain some semblance of standardisation. But how could Bond and Warder (or for that matter the various Regional engineers) be positive about the relative merits of what were in some cases as yet completely untried motive power designs? Luckily the Type 4 power range could already be considered as basically satisfactory at least for the diesel-electrics, because both the English Electric and the Sulzer engines had been well-proven in use in other countries. An initial recommendation was that the more powerful Sulzer engine should be adopted for all new Type 4 1Co-Co1s, but the greater construction capacity of English Electric had to be acknowledged. At one stage a modified design of 1Co-Co1, suitable for either make of engine to be fitted, as desired, was considered. Finally, production-version orders were placed for both classes, and in the case of the Sulzer engines the horsepower was increased from 2,300 to 2,500, with the EE Co engine remaining at 2,000hp.

A major consideration when replacing steam with diesel traction has to be the provision of suitable and clean maintenance and repair facilities and the training of all staff concerned with their operation. The original three-year plan had this worked out on the area by area basis. Now, with the sudden development of quantity production, the situation changed drastically. Drivers and firemen were required to learn the new skills whilst still also working on the steam footplate; whilst in many areas the expensive new diesels had to be housed and serviced alongside steam. The area dieselisation

Below:
The English Electric Co prototype No DP2 featured a modified 'Deltic' body fitted with a charge air cooled version of the 16SVT engine, producing 2,700hp. From 1962 until its demise in a collision near Thirsk in July 1967, DP2 performed excellently on both the West Coast and East Coast main lines, but BR instead preferred to order the untried Brush Type 4 design. *English Electric Co*

1S81
DP2
T

1Z18
HAWKER SIDDELEY
KESTREL
HS 4000

Left:
The 1.25pm Euston-Perth passes Stafford on 6 October 1962, hauled by DP2 around the Queensville curve. This locomotive formed the basis for the subsequent Class 50 design, and incorporated many sophisticated design features. *M. Mensing*

Below left:
The Brush/Hawker-Siddeley group expected high-output locomotives to be required in the near future and built a 4,000hp prototype numbered HS4000 and named *Kestrel*. Despite hauling some truly massive freight trains and successfully working some 'Deltic' diagrams on the ECML, *Kestrel's* potential for 125mph service was never unleashed and the idea of using high-power locomotives was abandoned in favour of the High Speed Train project. *IAL*

Right:
The cab design of *Kestrel* was produced to give maximum aerodynamic efficiency at high speeds, and the principal features are seen in this striking view of the locomotive at Cricklewood open day on 12 July 1969. BR operated *Kestrel* from 1968 to 1971, when it was returned to the manufacturer and subsequently exported to the USSR. *V. C. K. Allen*

Below:
The production 'Deltics' were visually far more attractive than the prototype version, and the lower band of pale green reduced the otherwise bulky appearance considerably. D9020 *Nimbus* enters Leeds Central with the down 'West Riding' in the 1960s. *E. Treacy*

Above:
'Deltics' at Doncaster on 12 May 1970, with No 9002 *The King's Own Yorkshire Light Infantry* arriving on an up express as No 9007 *Pinza* passes with the 14.00 King's Cross-Edinburgh. The full yellow cabs and all-blue bodywork gave the class a rather tubby appearance. *J. H. Cooper-Smith*

scheme — whilst not entirely abandoned — lost a great deal of credibility because there simply was not sufficient time to prepare in advance.

In the sections that follow I describe the initial allocations of the various batches of locomotives, and it will be seen that to begin with the LMR reaped the greater share of the initial Type 4 power. The E/NE Region had the 2,000hp EE Co vehicle as its most powerful diesel. This design, undeniably reliable and rugged, was not considered sufficiently powerful by the Regional authorities to meet future express passenger requirements and it was not long before pressures were being brought to bear upon the Chief Officers at Marylebone Road, for a more powerful design. Certain influential eyes were firmly focused upon the EE Co 3,300hp 'Deltic' — of which more anon — but there was also a positive case for producing a more powerful Type 4, with one uprated slow-running engine and electric transmission, or perhaps (taking a leaf out of Swindon's book) two quick-running German-type engines, but with electric rather than hydraulic transmission. The engine manufacturers themselves were looking at various ways of up-rating their products, and (an important factor) at ways of reducing the overall weight of the largest main line diesels. At the BTC headquarters the operating authorities were already looking for a Type 4 with fewer axles and greater route availability. If this could be allied to greater power, then so much the better.

At this stage some envious eyes were being cast towards the WR, where the Swindon version of the German B-B diesel-hydraulic was producing 2,200hp and 52,400lb of tractive effort with a total locomotive weight of only 79ton. Already the WR was thinking in terms of a C-C version with 2,700hp and a weight of about 105ton. Compared to the production version BR/Sulzer 'Peak' with 2,500hp and a weight of 138ton, the savings were obvious. Hydraulic transmission was however, already proving to have problems on the WR and the BTC were not keen to extend its use to other Regions. Another important factor was that British industry was more geared-up to the production of electrical transmissions, although some manufacturers had obtained licences to make German hydraulic transmissions and/or engines. By the early 1960s it was possible for the BTC to produce an outline idea of what the next generation of Type 4 diesel-electrics should look like (see p4). Thus the stage was set for the future, but there was another major character waiting unannounced in the wings! This was the EE Co 'Deltic', of course.

January 1959 witnessed the transfer of the prototype 'Deltic' from the LMR to the ER (see book one). At first it had to undergo the tedium of route clearance trials, but before long this formidable machine was demonstrating its true worth with test runs over the East Coast main line. To Mr G. F. Fiennes of the ER Board must go the accolade for having faith in this exceptional design and for pushing ahead the idea of a limited fleet of production 'Deltics' as future motive power for the principal East Coast main line express passenger duties. Behind this determination lay the knowledge that the LMR West Coast main line had priority for electrification. If the East Coast was to operate a comparable service, Fiennes realised that a locomotive capable of speeds of 100mph or more with good loads was an essential requirement. Early on in the pilot scheme days Fiennes had made his classic observation that 'The policy of building diesels of 2,000hp lies in ruins around us. Nothing less than well over 3,000hp under the bonnet will do'. Electrification of the East Coast route was still on the cards, but Fiennes witnessed Board meetings where the target date seemed to recede further and further. Meanwhile *something* had to be done in a hurry, if passengers and receipts were to be halted from their current decline. The 'Deltic' was to provide the solution.

The 'Deltic' was an expensive machine to build (and of course purchase) compared to the existing Type 4s and the envisaged fleet, which was supposed to be a 10-year stopgap pending electrification, would cost the ER some £4 million. There were also gloomy predictions about much increased maintenance costs. The ER wanted 23 locomotives,

Above:
Despite problems with the Napier two-stroke engines the class gave sterling service right up to the full HST era. No 9008 *The Green Howards* bellows away from Huntingdon with the 11.30 King's Cross-Leeds on 11 January 1971. *J. H. Cooper-Smith*

to replace 55 Pacifics, the BTC in their wisdom cut this to 22. For the production version the EE Co altered the overall dimensions of the bodyform to make it suitable for most major BR routes — no easy task with such a compact locomotive. The order was placed in 1958, and delivery was anticipated for some two years ahead. It must be a matter of lasting regret that the order was not increased to provide similar power for the Midland main line and perhaps the WR (although it was in all fairness too soon for the BTC to dismiss the hydraulics out of hand).

More about the 'Deltics' later, but first we must return to the outline specification for the second generation of Type 4 diesel-electrics. At the time this was being circulated there were at least three major private British locomotive builders who were keen to win orders for such a machine — the English Electric Co, the Brush Traction Co and the Birmingham Carriage & Wagon Co. Engine manufacturers with an eye to the future were EE Co, Sulzer and Bristol-Siddeley. BR collaborated with each of these firms as they prepared designs for prototype Co-Co Type 4 locomotives, for possible selection as the new BR standard. I intend to describe these prototypes in greater detail in a future book, but a brief description of each is necessary here in order to show the various possibilities BR had on offer at this time. Some 500 or more locomotives of one standard type were envisaged, of around 2,700hp with the body carried on two six-wheeled bogies with an axleload of 19 tons.

The BTC was abolished in the reorganisation of the 1962 Transport Act and Dr Beeching became Chairman of the new British Railways Board. At this time the post of Chief Mechanical and Electrical Engineer was held by Mr J. F. Harrison, who had taken over the job of both Bond and Warder combined. Harrison had earlier been instrumental in preparing the new Type 4 speci-

Right:
The Brush/Sulzer Type 4 was fitted with a 2,750hp version of the unusual twin-bank 12-cylinder engine fitted to the 'Peaks', but later modifications saw this unit derated to 2,580hp. No D1512 was one of the batch of 20 ordered straight from the drawing board. *BR*

fication, and in so doing he made a decision which was to have a profound effect upon the future appearance of the BR diesel fleet. This was that all new locomotives should have flat fronts rather than nose ends — a move away from the LMSR and EE Co-inspired appearance of the pilot scheme Type 4s. The BR Design Panel was very much in favour of this decision and it requested the firm of industrial designers, Wilkes & Ashmore, to prepare an initial idea of how the new locomotives should look. Wilkes and Ashmore had been the consultants for the pilot scheme Brush Type 2 and its BRCW counterpart, and more recently they had designed the appearance of the Beyer Peacock 'Hymek'. It was upon the style of this latter locomotive that they based the new front end layout, with a reduced height roof over the cabs at each end. Comparison with the prototype BRCW/Sulzer *Lion* design leaves no doubt that at one stage the BRB favoured this locomotive for its future standard, so exactly does the locomotive match the specification in terms of appearance! The exact reason why BRCW did not subsequently undertake to build for BR, and instead went into liquidation has an element of mystery surrounding it to this day.

The *Lion* with a single 2,750hp Sulzer engine, was finished in an incredible overall white livery, and was certainly a most striking design and an impressive locomotive in terms of performance. It appeared in the spring of 1962, and was in fact preceded by the rival Brush *Falcon*, a twin-engined machine of 2,800hp with two high-speed Maybach engines built under licence by Bristol-Siddeley. The *Falcon* was intended to demonstrate the virtues of high speed diesel engines when allied to electrical transmission. (At one stage BR actually considered producing a new Type 4 based upon the Swindon 'Western' class C-C, but using electrical transmission; these were to have been Nos D1200-1258 and would have been built at Crewe.) J. F. Harrison is on record as not greatly favouring high speed engines, however, and the *Falcon*, splendid machine though it undoubtedly was, started its career with this disadvantage. The original livery was an unusual combination of lime green and chestnut brown which was later changed to the standard two-tone green of BR. Wilkes & Ashmore once again advised Brush Traction upon the external appearance and the front end was evolved from the Brush Type 2 design, but without gangway doors and with only two cab windows in the front.

The BRCW *Lion* already referred to, matched BR's specification even to the 19-ton axleload and it had the most powerful single engine then available, being Sulzer's latest version of their twin-bank engine. It was in many respects a most sophisticated machine, with features such as dual train-heating (steam and electric), an Alsthom rubber bogie/superstructure suspension and a stressed-skin steel lightweight body construction. Up to waist level the design consultants advised a series of flutes along the bodyside to improve the finished appearance. All louvres were above cantrail level. It took a long time for BRCW to build *Lion*, possibly too long for BR's liking, and the firm was in serious difficulties during this period. Perhaps we will never know the full story, but it seems that BR decided to hedge its own bet, and a second Sulzer-engined Co-Co design was put in hand by Brush Traction (who possibly foresaw that *Falcon* would not be accepted as the new BR standard). Of this, more anon; but first we must briefly look at the third of the privately produced Type 4 prototypes. This was the EE Co 2,700hp Co-Co, No DP2

Right:
The first 20 Brush Type 4s were fitted with an electrical train-heating generator, and of these all apart from No D1500 were also fitted with steam-heating boilers. The pioneer locomotive is seen here running as No 47401 and named *North Eastern*, hauling the 09.03 Liverpool-Newcastle away from Leeds on 20 March 1987. *S. R. Batty*

Below right:
Experience gained with DP2 was put to good use in the production of 50 2,700hp Co-Co locomotives initially intended for use on the West Coast main line prior to full electrification. Nos 404 and 431 depart from Crewe with the 08.05 Euston-Glasgow on 22 March 1971. *J. H. Cooper-Smith*

Below:
Known in post-TOPS days as Class 47, the Brush design has penetrated to all corners of BR and worked all varieties of traffic. No 47148 passes Goose Hill Junction, Normanton, with a ballast train on 3 August 1986. *S. R. Batty*

47 401

1S47

4M55

Left:
Class 50 No 425 passes Oxenholme with an up Freightliner on 18 May 1971. The BR-specified cab headcode panel and untidy jumper cables spoiled what could have been a very neat cab-front appearance. *J. H. Cooper-Smith*

which appeared in 1962. Externally this excellent locomotive did not conform to Harrison's outline specification, being in fact basically the same as the production 'Deltics', except for some modified louvres. Finished originally in an overall dark green, with a grey roof and black bogies, and carrying the by then obligatory yellow warning panels at each end, No DP2 (for some reason it was not favoured with a name) was the only prototype to feature a British designed and built engine.

No doubt the use of the existing 'Deltic' bodyform saved considerably upon production costs, but it was against Harrison's expressed preference for flat-fronted cabs. No DP2 proved to be a magnificent performer on the road, but unfortunately it came just too late for BR's urgent needs, and even whilst all three prototypes were still running trials, BR placed an order for 20 new 2,750hp Co-Cos with Brush (as already mentioned), to a design as yet untried and untested. This was of course the classic D1500, or Class 47 as they are now known.

Outwardly D1500 followed the Wilkes & Ashmore bodyform of *Falcon* and *Lion*, although perhaps less successfully than either of these. An attempt to keep weight down resulted in the use of very large oleo-pneumatic buffers, instead of the intended oval-shape. The original two-tone green livery with the yellow panel integrated with the colours was certainly striking and did much to relieve the blank bodyside panels, but it is a pity that the fluting used on *Lion* was not repeated. No D1500 featured the (uprated) Sulzer engine and electrical equipment of the 'Peaks', but in many other respects it was a very different machine indeed from both the 'Peaks' and *Lion*. BR advised Brush on which features of *Lion* they favoured, and it seems now that they were seeking a less sophisticated, but very rugged machine capable of the whole range of mixed traffic duties. A total of 512 locomotives was eventually built to this design (with various detail modifications as described in Section 4) and today's rail fleet still largely depends upon these versatile machines.

By the mid-1960s the BRB was intent upon rationalising its diesel locomotive fleet, to reduce both the variations of engine and equipment, and the number of locomotives required. Known as the 'National Traction Plan' the BRB produced a confidential report upon the design and availability of all the diesel classes it possessed. This report was admirably unbiased and it led to the early withdrawal of some diesel locomotive classes, even including locomotives of recent manufacture. The Brush Type 4 did not emerge unscathed from this searching examination and there were detailed criticisms of the bogie design, the engine and transmission and the overall reliability. Nevertheless by this time these Co-Cos had proved their worth, and it was considered worthwhile to set about remedying their troubles and extending their service life.

A desire to increase operating speeds, for commercial gain, was also being voiced by the mid-1960s. The 'Deltics' were showing off their paces on the East Coast main line, but they were also proving to be exceptionally expensive locomotives to maintain. Within EE Co itself there was some resistance to further development of the 'Deltic' by people who preferred to see their own medium speed engine developed further (as it had been for No DP2). By the autumn of 1966, however, the 'Deltic' protagonists had produced an outline scheme for a 4,400hp version which they dubbed the 'Super Deltic'. This would have had a 19-ton axleload, and the BRB flat front cab layout. It came to nothing because one of its greatest supporters on BR, Mr G. F. Fiennes (who would have pressed the case for construction) was peremptorily dismissed from his post at the BRB after publication of his book *I Tried to Run a Railway!*

Sulzer's were also interested in producing a more powerful high speed diesel locomotive using a development of their LVA24 engine. BR themselves were beginning to think in terms of a high-powered locomotive suitable for express passenger and heavy freight duties and they produced an outline specification for a Type 5 Co-Co which must not exceed 126 tons in working order; in 1964/65. Sulzer and Brush then collaborated to produce the *Kestrel* as a private venture, completed at Loughborough in November 1967. This impressive machine will be described in detail later in this series. It unfortunately proved to be somewhat heavier than the stipulated weight. The design consultants Wilkes and Ashmore were again called upon to advise on the appearance of the locomotive, and in view of the envisaged higher speeds they gave the cabs an aerodynamic form considerably more streamlined than the Brush Type 4. The livery was a startling combination of chocolate brown to waist level and golden yellow above. This 'might-have-been' Type 5 was not finally taken up by BR, whose traction policies had meanwhile undergone a fundamental change of heart.

The BRB had concluded that for future construction the two main categories of train operation should be met by separate designs. For high-speed train operation special locomotives or multiple-units would be required, whilst a new generation of heavy duty freight locomotives would become necessary in due course of time. However, before this new policy could be initiated in design terms, there was an immediate requirement for 50 more Type 4 locomotives of around 2,700hp, for use on the LMR West Coast main line, pending electrification through to Glasgow from Weaver Junction. Once electrification was completed it was intended to transfer all 50 to the WR to replace the 'Western' class diesel-hydraulics. To match the excellent performance the electric locomotives were providing at the southern end of the Anglo-Scottish services, the new locomotives would be required to work in pairs, in multiple; thereby allowing an improved timetable to be prepared in advance of electrification. But what sort of locomotive was to be ordered?

The proposed 'Super-Deltic' could have filled the bill admirably, of course and

Below:
Several of the Class 50's sophisticated features caused many headaches for running staff, and were removed during a refurbishment programme undertaken by BREL Doncaster. One of the modified examples is seen climbing Dainton Bank with a down West of England working in the summer of 1981. Notice the blanked-off headcode panel, black cab window surrounds and headlight. *S. R. Batty*

without double-heading, but it was not a truly proven machine. And then again, with 512 Brush Type 4s in service it might have seemed reasonable simply to build a further 50 of these, as both the Regions concerned had many locomotives of this type already. But no: the eyes of the BRB were trained upon the performance that the EE Co prototype Type 4 No DP2 had been displaying. (Sadly it was shortly afterwards to be damaged beyond economic repair in an accident near Thirsk.) It was considered to be in many respects the most successful diesel-electric design operated in Britain to that time, and the idea of a fleet of 50 such locomotives had much attraction, particularly as the Brush/Sulzer Type 4s were still proving troublesome in certain respects.

The new D400 class, today's Class 50, was a revised version of No DP2, featuring the flat-fronted cabs requested by BR and with a number of additional and sophisticated devices, which resulted in a cramped internal layout, complete with a host of in-built problems. With hindsight BR would perhaps have been better off to have purchased 50 more DP2s in original and simple form! However today, after much rectification and bitter experience, the Class 50s have been moulded into shape.

The decision taken in 1968 to build High Speed Trains with multiple-unit power cars sandwiching Mk 3 carriages, and to continue development of the more sophisticated Advanced Passenger Train, meant that the era of the high speed mixed traffic diesel-electric locomotive was over. No more Class 50s could be justified, but there was a new and fast-growing need for more heavy freight locomotives by the early 1970s. In 1974 an order was placed with Brush for 30 new 3,250hp Co-Cos; to be BR Class 56. This order followed much speculation as to what sort of locomotive should be produced. At one stage the idea was seriously considered of using the power cars of the proposed High Speed Trains, minus the carriages of course, coupled together and working (with a suitably varied gear ratio selection) on freights during the night. A basic feature of the HSTs is that they are not sleeping car trains. There were various twin-engined Co-Co units in the 4,000-6,000hp range proposed, and a 25kV electric Co-Co was envisaged using identical body, bogies and traction equipment. However, what finally emerged in answer to an urgent demand was the Class 56 with Brush electrical equipment, English Electric engine and a body form clearly derived from the Brush Type 4. The EE Co engine is a direct descendant of that used in LMSR units Nos 10000/1, but producing *twice* the power! Such is the development work that engineers have now successfully put into practice.

After some initial problems, described in Section 6, the Class 56 has settled down well in everyday service. Originally the class was due to total 170 examples, but construction ceased with No 56135, and a start was made instead on the new Class 58 fleet in 1982.

The Class 58 was intended to replace the quickly-produced Class 56s and was envisaged at the time as being the true 'next generation' freight locomotive. Relatively cheap to produce due to simple and rugged construction, great hopes were held that an export market would be found for the class, but this has not transpired. A fleet of 150 Class 58 locomotives and no fewer than 200 of the 25kV electric version — Class 88 — was at first planned, but only 50 of the diesel version, and none of Class 88, were produced.

The years 1980/81 saw a severe recession in BR's traffic (freight in particular) and many locomotives were withdrawn as surplus to needs. The better examples were stored at Swindon works pending a return to use (and were officially described as 'withdrawn from operating stock'), but several classes never recovered from this 'mothballing' operation. Class 44 (Pilot Scheme) 'Peaks' disappeared, and their Brush-built Class 46 brethren disappeared during 1984, and the last main line English Electric Class 40 locomotive followed early in 1985, except for the first of the class, No D200/40122, retained for use on special trains until 16 April 1988. The remaining Class 55 'Deltics' in traffic, Nos 55002/09/15 and 22 were finally withdrawn from traffic on 2 January 1982 because of their displacement by HST sets and also because of their very high maintenance costs. In more recent years the significant events have been the crippling losses suffered to freight traffic during the miners' strike (1984/85), the arrival of the 'Sprinter' DMU fleet and subsequent dispersal of locomotives (1986), the resurgence of Railfreight business (1987) and the genesis of new freight locomotives of Class 60 and perhaps Class 65 (1989). On the debit side, the Class 45/1 'Peaks' have disappeared far earlier than many expected, several Class 47 locomotives

Above:
Such was the hurry to produce a heavy freight locomotive for use in the late 1970s that the first 30 examples of Class 56 were built under licence in Romania. One such example, No 56011, is seen in original condition hauling a load of coal towards Ferrybridge on 22 December 1986. The small badges and numerals placed on such a vast expanse of blue paint created a very dull-looking locomotive. *S. R. Batty*

Right:
Class 56 No 56074 *Kellingley Colliery* works hard past the site of Mirfield depot with a load of coal bound for Fiddlers Ferry power station on 27 October 1982. The warning beacon above the cab was fitted as part of trials in remote control of the locomotive during loading and unloading of the train.
A. V. Haymes

have been withdrawn and the Class 50 fleet appears to have only a short life ahead.

As I have stated in the first book of this series, it took BR many years and many heartaches before it achieved today's main line diesel-electric fleet. No single class of locomotive has entered traffic without problems, and even the Class 58 design had its share of headaches and teething troubles, before settling down to prove its true worth.

Engine manufacturers have had to combat unsuspected faults and weaknesses revealed by the rough treatment of everyday BR operations, and some deratings either temporary or permanent, have proved necessary. Refurbishing has enabled the life expectancy of certain classes to be radically improved, in particular the Brush/Sulzer Class 47s, which went through a very rough patch. Central policy decisions, such as the adoption of the air brake instead of vacuum, the fitting of BR AWS equipment and the changeover of train heating from steam to electricity, have all required locomotives to be modified accordingly — sometimes with precious little space to spare for new items of equipment.

In presenting this survey of today's BR Type 4 and 5 diesel-electric fleet I am deeply conscious of the tremendous effort that has gone into its creation by countless people, at all levels of both the BR and the private industry sectors. It would be invidious to mention particular personalities, as we are dealing not just with recent history but, of course, with the present day and the future. My sincere thanks, however, to many people who have assisted in its compilation.

To any newcomers to this series, I should explain that the main objective is to present in broad chronological sequence a picture of the growth and origins of today's BR fleet. In books 1 and 2, I have described the early prototypes and the pilot scheme designs, and the 20-year saga of the use of hydraulic transmissions on the WR. The next volume describes the production series Types 1, 2 and 3 diesel-electrics.

In the section that follows, I have tried to create a picture of the main changes, of livery and various details that have affected the production Types 4 and 5. It is not intended to be a technical treatise, and the reader in search of more detailed information on this aspect is recommended to the following books, which collectively present an excellent survey.

Deltics at Work; Allan Baker and Gavin Morrison; Ian Allan Ltd.
Last Days of the Deltics; C. Fellows and P. Gash; Ian Allan Ltd.
The Deltics. A Symposium; Cecil J. Allen, G. F. Fiennes, etc, etc; Ian Allan Ltd.
Class 47 Diesels; A. T. H. Tayler, W. G. F. Thorley and T. J. Hill; Ian Allan Ltd.
Locomotive Recognition: Class 50s; Colin J. Marsden; Ian Allan Ltd.
Model Railway Constructor Planbook: BR Main Line Diesels in 4mm Scale; Russell S. Carter; Ian Allan Ltd
English Electric Main Line Diesel Locomotives of British Rail and *Sulzer Diesel Locomotives of British Rail*; both Brian Webb; David & Charles.
Brush Diesel Locomotives 1940-1978; George Toms; TPC/Turntable.

The railway modeller is very much in mind in my presentation herein, and I hope the subtleties of change which each class has undergone and which I have described, will be of assistance in giving greater accuracy to models of their prototypes. The reader in search of an accurate list of the present-day fleet is recommended to the current edition of

the Ian Allan ABC British Rail Locomotives. In addition, each month detailed stock changes are listed in the Ian Allan periodicals, *Modern Railways*, *Motive Power Monthly* and *Railway World*.

Finally, as this is a series which, it is anticipated will span some considerable period of time, and one which has involved the author in a great many hours of research and compilation, there is room in future books for any important addenda. Therefore, any reader who can contribute further useful information to any relevant aspect, or who detects an error or omission on my part, is invited to contact the author, c/o Ian Allan Ltd, Terminal House, Shepperton, Middlesex TW17 8AS. I regret I cannot enter into specific correspondence, but all contributions will be gratefully acknowledged.

Above:
Many of Class 56 were painted when new in the 'Large Logo' style including full yellow cabs and black window surrounds. No 56133 *Crewe Locomotive Works* passes Marsden on the descent to Huddersfield with a Glazebrook-Haverton Hill petrochemical train. *C. J. Tuffs*

Left:
Although hailed as the 'next generation' of freight locomotive at its introduction, the Class 58 has not been without its troubles and no further building is planned. No 58033 approaches Stenson Junction with the 08.35 Garston Docks-Mansfield empties as sister No 58009 eases her train through the track hopper house at Willington power station on 7 January 1986. *A. O. Wynn*

Absent Friends

During recent years Classes 40, 45 and 55 have disappeared from the BR main line traction scene. This short pictorial feature is intended as a reminder of the activities of these robust and well-liked locomotives which formed such a familiar part of the railway scene for so many years.

Left:
The English Electric Class 40 1Co-Co1 locomotives were regular performers on the Newcastle-Leeds-Manchester-Liverpool service until the 'Peaks' almost made a complete takeover. No D280 is seen between Marsden and Slaithwaite with the 15.00 Liverpool-Newcastle on 22 June 1971.
J. H. Cooper-Smith

Below:
A North Wales-Euston train passes Chester No 2 signalbox behind Class 40 No 319 on 16 June 1972. The length of these locomotives tended to reduce the awful visual effect of the 'Rail blue' paint! *C. Plant*

IV 77

Above left:
Class 40 No 40046 heads along the Calder Valley with a westbound freight on 21 April 1982. Notice the welded-up gangway doors and royal train headcode! *L. A. Nixon*

Left:
The 'Peaks' regularly reached Western Region territory, especially after the demise of the diesel-hydraulics. No 45109 passes Dawlish with a Nottingham-Paignton train on 11 August 1973. *N. E. Preedy*

Above:
With split headcode boxes still in use, No 45017 grinds up the Lickey Incline with the northbound 'Cornishman' on 8 October 1974. *P. D. Hawkins*

Right:
The 15.10 Liverpool-Newcastle departs from Lime Street on 16 April 1976 behind Class 45/0 No 45053. *P. Gerald*

1S35

8n
167
1S32

Left:
'Deltic' No 9017 *The Durham Light Infantry* passes Hatfield with the 14.00 King's Cross-Edinburgh on 13 February 1970. These remarkable machines saw the ECML through the transformation from steam to HST power. *J. H. Cooper-Smith*

Right:
No 9007 *Pinza* accelerates away from King's Cross with the 17.05 for Leeds on 31 July 1970, leaving a trail of exhaust which would have done credit to any Gresley Pacific!
J. H. Cooper-Smith

Below:
The first production 'Deltic', No 9001 *St Paddy*, gets away from York with the 12.00 King's Cross-Edinburgh on 8 September 1971. *J. H. Cooper-Smith*

Pilot Scheme Designs Perpetuated

The headlong plunge into quantity production of new diesel locomotives, which effectively curtailed the BR pilot scheme, was responsible for the perpetuation of both the Type 4 1Co-Co1 designs. It is probable that, had the three-year evaluation period been fulfilled, neither would have gone into quantity production without change. In particular the heavyweight bogie design, inherited from Bulleid's SR diesels was the subject of some concern to the civil engineers. Although they produced a good ride for the locomotives they were hard upon the body structure and equipment and on the track. Tight curves were a particular hazard, in shed and goods yards; whilst the long wheelbase needed to provide an axle loading of 18-tons prevented the locomotives from passing over the hump in marshalling yards. Nevertheless, both designs have given sterling service to BR, although the EE Co design which remained rated at 2,000hp (whereas the Sulzers were uprated to 2,500hp) has probably been the more reliable overall. The EE Co version, Class 40, disappeared from main line use in 1985, except for No D200/40122, retained for special duties, and the last departmental survivor went in 1987; the Brush/Sulzer 'Peaks' have all gone, even the ETH-fitted Class 45/1 variety being swept away during 1988, long before their usefulness had expired.

Below:
The production EE Co Type 4 locomotives were identical in appearance to the pilot scheme batch, having Brunswick green livery and grey roof separated by a blue-grey stripe. Gangway doors were also fitted, but the yellow warning panel was a later addition. No D315 passes Scout Green box on the climb to Shap Summit with the 10.40 Euston-Carlisle in July 1963. *M. Pope*

1 English Electric Co 2,000hp 1Co-Co1 Type 4

Class: BR Type 4, later Class 40
Purpose: Heavy duty locomotives
Introduced: 1958
Original numbers: D200-D209 pilot scheme; D210-D399 production series
Total built: 200 (10 pilot scheme, 190 production)

Details of Nos D210-D399
Engine: English Electric 16 SVT Mk II
No of cylinders and cycle — 16 four-stroke
Max cont rated output — 2,000hp at 850rpm
Main generator: English Electric Type EE82
Traction motors: Six English Electric EE526/5D or 526/7D
Type of suspension — nose
Type of gear drive — single reduction
Performance: Max tractive effort — 52,000lb at 21.0% adhesion at 2,700A main generator
Cont tractive effort — 30,900lb at 18.8mph at 1,800A main generator
Rail hp at cont rating — 1,550hp
Full engine output — available between 10 and 79mph

Braking: Straight air and auto air for loco, auto air and air cont vacuum for train giving a brake force of 63.5% of loco weight in working order
Speed: 90mph (max permitted service)
Curve: 4.5 chains min rad curve with gauge 0.75 widening, 3.6 chains horizontal with 0.75in gauge widening
Train heating equipment: Clayton RO 2500 Mk 1 giving a steaming capacity of 2,500lb/h
Tank capacities: Engine fuel — 730gal
Boiler fuel — 200gal or 170gal (D260-D399)
Boiler water — 800gal

The English Electric Co did not need to make any significant major modifications to their Type 4 design, when production orders were placed by the BTC. The 10 pilot scheme locomotives (see Section 8 in Book one of

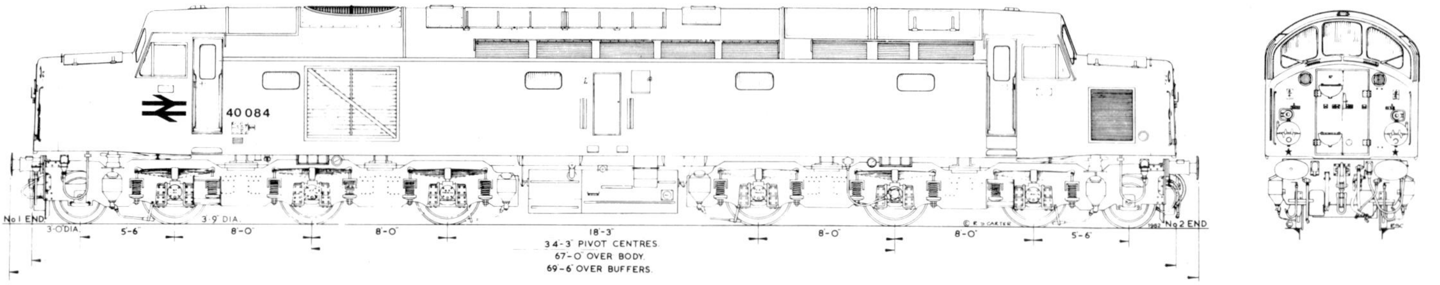

Below:
2,000hp 1Co-Co1 English Electric Co Ltd mixed traffic Type 4; a direct production follow-on from pilot scheme Nos D200-D209. The drawing shows the earlier nose end variant with gangway doors. *R. S. Carter*

Bottom:
The EE Co Type 4s took over from steam power on the Trans-Pennine Northern route before the 'Peaks' arrived in large numbers. No D271 approaches Diggle station with the 15.00 Liverpool-Newcastle on 27 July 1966. *J. Clarke*

this series) had given a good account of themselves and delivery of the first production batch, for use on the LMR West Coast route commenced in May 1959 (Nos D210-D236). The following year the LMR gave some consideration to up-rating these locomotives to 2,400hp by means of charge-air cooling; but this was not put into practice, and the entire class have spent their lives with their original 2,000hp rating.

The cumbersome bogie design, derived from Bulleid's SR 1Co-Co1s, gave some trouble in service with fires and fractures, and modified springing was introduced to reduce stresses, but not before the hard ride given by these bogies had (due to vibration) caused problems with traction motor flashovers at high speeds. On the LMR a programme of replacement using four-pole traction motors instead of six-pole solved this problem. The original six-pole motors were reused on new EE Co Type 1 D8000 series Bo-Bos.

Left:
Class 40 No 217 *Carinthia* passes Holywell Junction with the 10.15 Euston-Bangor on 29 March 1973. Fortunately the full yellow treatment was confined to the nose front, and cab windows and bonnet sides were spared.
A. W. Hobson

Below:
The headcode discs remained on those locomotives so fitted throughout their lives. No 40074 passes Aberdeen Ferryhill with the 08.10 Glasgow-Aberdeen on 21 March 1974.
B. Morrison

Left:
Several locomotives were built with split headcode boxes placed either side of the gangway doors. These were No D255 (also fitted with electrical train heating — see text) and Nos D324-D344. In this scene, the former No D336 is seen as 40136 hauling a container train towards Portobello depot through Edinburgh Princes Street Gardens on 4 May 1979. *G. A. Watt*

The production batches, six in all, went to the LMR, ER, NER, and ScR. One odd-man-out was No D255 which was experimentally fitted with electrical train heating and an anti-slip device and allocated new to Derby for trial purposes. In due course the class was ousted from the premier express passenger duties, either by electric traction (on the LMR and ScR) or by more powerful diesel locomotives. Air brake equipment added for train working and lack of boilers (with no boiler water or fuel carried) meant that the weights of individual locomotives varied somewhat.

In the early days of dieselisation the Class 40s, proved to be one of the best in the Type 4 power range, despite being the least powerful. They gave excellent service on both East and West Coast main line passenger duties, taking on the duties previously tackled by Gresley, Thompson and Stanier Pacifics. In so doing they suffered the rigours of sharing shed facilities with steam, and maintenance in steam surroundings was unavoidable for some years; a situation that was far from ideal. Withdrawal of the class was largely completed during January 1985, and four examples were taken for Departmental use (sometimes from the scrapline, and only a hair's breadth from actual cutting) later during that year. Nos 40012/060/118/135 became Nos 97405-408 respectively and were used principally for

Above:
Class 40 No 40034 runs light engine into York from the south on 23 September 1976. This locomotive has acquired a coat of yellow paint on the bonnet top, but not on the access doors. *P. D. Hawkins*

Below:
After gangway doors were finally abolished, a much neater central headcode panel was fitted. No 40152 emerges from Standedge Tunnel with the 10.10 Liverpool-Newcastle on 8 July 1976. The locomotive number has been moved back on to the bodyside and the BR symbol placed on the cabside. *D. A. Flitcroft*

Above:
Looking rather neglected during her last days, No 40044 passes Milford Junction and heads towards York with a train of empty container flats on 1 March 1984. Her paintwork is rather battered, and the outer grille to the bodyside louvres is missing. *S. R. Batty*

work connected with the Crewe station rebuilding, although they were frequently seen at work on engineer's trains throughout the northwest and occasionally worked the odd passenger or freight duty. These locomotives enjoyed only a briefly-extended life with BR, and after ending their days in the Liverpool area, No 97406 (ex-40060) was withdrawn as the last survivor during 1987.

Happily, this was not the end of the story for these popular machines. No 40106 had been used extensively for railtours, and this green-liveried example was bought for preservation on the Great Central Railway at Loughborough. After withdrawal, its place on the railtour scene was taken by No 40122, similarly repainted and also renumbered at one end with its 1958 number of D200. This locomotive was refurbished at Toton depot during the summer of 1983, using a power unit and one bogie from No 40076, and there then followed almost five years of use on railtours and regular passenger duties for this extremely popular locomotive. After a Liverpool Street-Norwich run to commemorate the 30th anniversary of the line's dieselisation, which had also been inaugurated by this very same machine, No D200 arrived at the National Railway Museum under her own power on 16 April 1988 for permanent exhibition. At the time of writing five further examples have either been restored or are due for preservation. No 40012 has been moved to Butterley, No 40013 to Meadowhall, in Sheffield, and No 40118 has been put on display in the Birmingham Railway Museum. The East Lancashire Railway, at Bury, have acquired No 40135 by courtesy of the Class 40 Preservation Society, where No 40145 — the solitary preserved Class 40 with a centre headcode panel — will soon be restored to full running order.

Below:
Class 40 No 40174 throbs over Ais Gill Summit with a diverted train of empty newspaper vans on 6 May 1984. Today, the Class 40s have gone and the newspaper traffic has been lost to road competition.
S. R. Batty

Below right:
As related in the text, several Class 40s survived a little longer as Departmental locomotives, being used on relaying work but occasionally being pressed into revenue-earning duty. The former No 40060, renumbered to 97405, shunts an engineers' train at Preston on 20 September 1985.
S. R. Batty

2 British Railways/Sulzer 'Peak' 2,500hp 1Co-Co1 Type 4

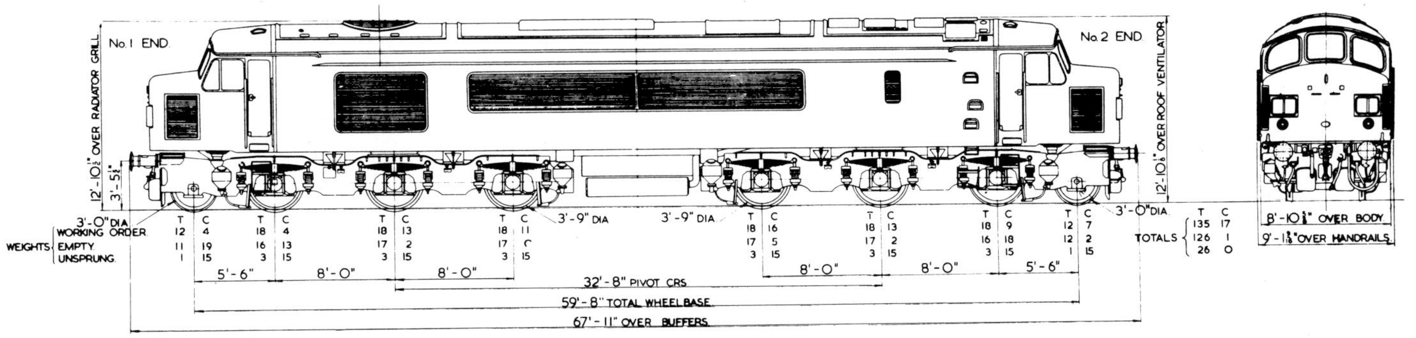

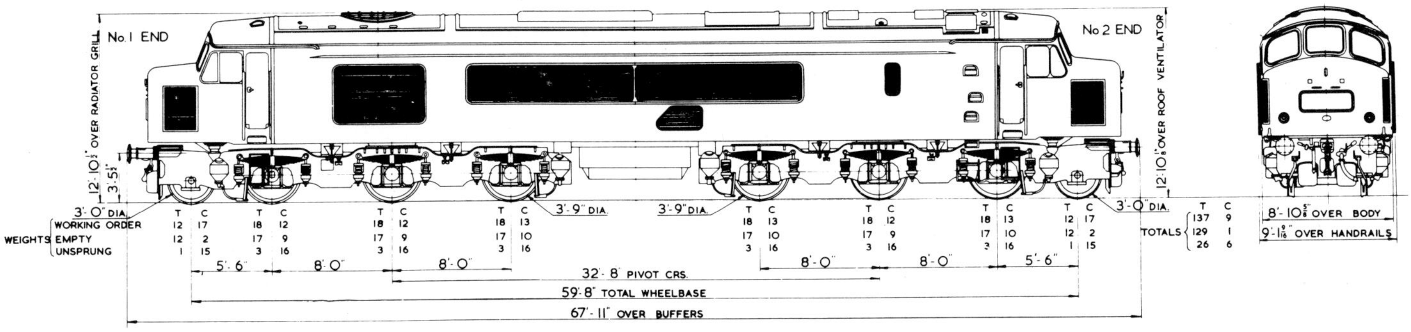

Above:
Production versions of the BR/Sulzer 'Peak' class 1Co-Co1, with engine uprated to 2,500hp. Top: Sulzer/Crompton-Parkinson version (Class 45). Bottom: Sulzer-Brush version (Class 46). Two headcode panel variations depicted; both after the decision not to perpetuate the central gangway doors. (All the Class 46 were built new with the single central four panel route indicator.)

Class: BR Type 4, later Class 44*/45†/46●
Purpose: Mixed traffic main line locomotives
Introduced: 1959
Original numbers: D1-D10* pilot scheme; D11-D137†, D138-D193● production series
Total built: 193 (10 pilot scheme, 183 production)

Details of Nos D11-D137† as built
Engine: Sulzer 12LD28B
No of cylinders and cycle — 12 four-stroke
Max cont rated output — 2,500hp at 750rpm
Main generator: Crompton Parkinson CG426A1
Traction motors: Six Crompton Parkinson C172A1
Type of suspension — nose
Type of gear drive — straight spur
Performance: Max tractive effort — 55,000lb at 22.1% adhesion at 4,200A main generator
Cont tractive effort — 30,000lb at 25mph at 2,670A main generator
Rail hp at cont rating — 2,000hp
Full engine output — available between 12 and 82mph
Braking: Air for loco, vacuum for train giving brake force of 78% of loco weight in working order
Speed: 90mph (max permitted service)
Curve: 5 chains (min rad curve without gauge widening at dead slow speed)

Above:
A rather grimy No D92 passes Halton with the 12.30 Morecambe (Promenade)-Leeds (City) on 15 April 1963. The grey roof and yellow warning panel are plastered with filth, but the off-white bodyside band is visible. Split headcode boxes continued to be fitted for some time after plain cab fronts (ie without any gangway doors) became standard.
N. A. Machell

Train heating equipment: Stone Vapor OK4625 boiler with 2,750lb/h steaming capacity
Tank capacities: Engine and boiler fuel — 840gal
Boiler water — 1,340gal

Details of D138-193●
Engine: Sulzer 12LDA28B
No of cylinders and cycle — 12 four-stroke
Max cont rated output — 2,500hp at 750rpm
Main generator: Brush EE Co TM73-68 Mk III
Type of suspension — nose
Type of gear drive — straight spur
Performance: Max tractive effort — 55,000lb at 22% adhesion at 3,30A main generator
Cont tractive effort — 31,600lb at 23.3mph at 2,100A main generator
Rail hp at cont rating — 1,962hp
Full engine output — available between 11.2 and 85mph
Braking: Air for loco, vacuum for train giving brake force of 77.1% of loco weight in working order
Speed: 90mph (max permitted service)
Curve: 5 chains (min rad horizontal without gauge widening)

Above:
Later 'Peaks' carried a centrally-placed four-character headcode panel arrangement as shown in this view of No D126 leaving Leeds with the up 'Thames-Clyde Express' on 18 May 1967. Under TOPS, the 'Peaks' were not renumbered in numerical order, and several apparently younger examples could be seen carrying older cab-front designs.
L. A. Nixon

Left:
The Midland main line was an early 'Peak' stronghold, and No D129 is seen sweeping along through Kibworth with an up express in freezing conditions on 10 January 1968.
J. H. Cooper-Smith

Below left:
'Peaks' were the first large diesel-electrics to reach the WR, finding their way to Devon and Cornwall on Northeast-Southwest duties. D89 *Honourable Artillery Company* nears Ivybridge with the 13.40 Plymouth-Leeds on 27 August 1971. *J. H. Cooper-Smith*

465 chains (min rad horizontal with 0.75in gauge widening)
23 chains (min rad vertical convex)
24 chains (main rad vertical concave)
Train heating equipment: Spanner Swirlyflo Mk III boiler with 2,750lb/h steaming capacity
Tank capacities: Engine and boiler fuel — 835gal
Boiler water — 1,040gal

The initial pilot scheme batch of 10 'Peak' class 2,300hp Type 4s are described in Section 15 of Book one of this series. One of these locomotives, No D2 *Helvellyn* was experimentally fitted with a charge air cooler for some months and the engine uprated to 2,500hp. This was successful enough to allow the higher rating to be used for the production orders. The 10 original 2,300hp

Above left:
The Midland main line was a regular haunt of the class, even beyond the introduction of HSTs in 1982 and right up to occasional workings in 1988. Class 45 No 156 emerges from Duffield Tunnel with the 08.35 St Pancras-Manchester on 22 October 1971.
J. H. Cooper-Smith

Above:
An up Sheffield express passes Corby behind No 62 *5th Royal Inniskilling Dragoon Guards* on 30 April 1972. The application of rail blue paint saw the end of the off-white body stripe, and the nose-ends were painted full yellow.
J. H. Cooper-Smith

Left:
A total of 37 'Peaks' received names, including the 10 Pilot Scheme Class 44 locomotives. Many were taken from withdrawn 'Royal Scot' and 'Patriot' steam locomotives, and the names were bestowed upon the Class 45/46 diesels in a haphazard fashion between September 1961 and October 1967. No 68 *Royal Fusilier* is seen at Saltley depot on 6 March 1974, and the plated-over steps to the original boiler water filling point can be clearly seen — this was transferred to solebar level to avoid danger to staff in areas with overhead electrification equipment. *P. D. Hawkins*

locomotives had Swiss-built engines, but for the production series the 2,500hp Sulzer engines were built by Vickers Armstrong Ltd at their Barrow factory, under licence. An order for 137 locomotives (later reduced by 10) was placed by the BTC, with Crompton Parkinson electrical equipment. A further order, for 76 locomotives featured Brush electrical equipment instead. Of these 76, the last 20 were not constructed as 'Peak' class 1Co-Co1s but appeared instead as Nos D1500-D1519, built by Brush (using much equipment ordered for the 'Peaks') and these were, of course, the first examples of the BR standard Type 4 Co-Co. (See Section 4.)

Both Derby and Crewe works built the production 'Peaks' as follows: Nos D11-D49/ D138-D193 at Derby and Nos D50-D137 at Crewe. Externally the production locomotives differed but little from the pilot scheme design, except at the nose ends where headcode panels replaced the disc and electric light headcodes. To begin with the gangway connection was retained in the nose, necessitating the four characters to be grouped in two pairs, on each side. The gangway connection was, however, soon abandoned and later locomotives were built new without this feature. The livery for the production series was the same as that devised for Nos D1-D10. Names were not applied when built, but starting with No D100 *Sherwood Forester* in 1961, a total of 26 of the class have since been named.

As in the case of the EE Co Type 4 1Co-Co1s, the SR/Bulleid bogie design has given a certain amount of trouble, now rectified and for a period there was a recurrent problem of bogie fires and fractures. More serious trouble arose, on the Crompton Parkinson locomotives, Nos D11-D137, with their traction motors and this affected their availability for a considerable period of time. On the Brush-equipped locomotives, Nos D138-D193 there was trouble with condensation in the control cubicle. The basic reason for changing from Crompton Parkinson to Brush electrical equipment for these locomotives was that the former firm was already so heavily engaged in production that it could not offer early delivery to BR. Under the BR reclassification scheme Nos D11-D137 became Nos 45001-45077 and 45101-45150 of Class 45, whilst the Brush-equipped locomotives became Nos 46001-46056 of Class 46. The series commencing 45101 onwards were locomotives fitted with electrical train heating, to work modern passenger rolling stock. The ETH control cabinet occupied the space formerly housing the boiler and the water tanks were removed and replaced by ballast weights and repositioned auxiliary apparatus, including the firebottles and toilet water tank.

For the period May 1962 until the spring of 1964 'Peak' class 1Co-Co1 No D57 ran with an uprated engine, providing 2,750hp. This was the forerunner of the engines used in the Brush Type 4 Co-Cos, today's Class 47.

The 'Peaks' were always associated predominantly with the Midland main lines and the cross-country services via Birmingham, from the northeast to the southwest. They were the first large diesel-electrics to be received by the WR, and they soon penetrated the diesel-hydraulic stronghold of Devon and Cornwall on these services. Some of the Brush-equipped batch went new to the NER at Gateshead for East Coast main line service as well as Trans-Pennine duties, such as Newcastle-Liverpool. The Class 46 locomotives ended their days allocated to the Eastern Region, and all had gone by the end of November 1984. Four were allocated to Departmental duties when Nos 46009/23/35/ 45 became 97401-4 respectively. Only Nos 97403/4 survive today, at the Derby research establishment; No 97402 was withdrawn during 1984, and 97401 suffered a tragically spectacular end during a 100mph collision test with a nuclear fuel flask in July 1984.

Withdrawals of the '45/0' series commenced during 1980 and the first Class 45/1 locomotive was taken out of service after severe collision damage in January 1985. The summer of 1986 saw the 'Peaks' have their last fling on the North Trans-Pennine route prior to displacement by Class 47s and the much-vaunted but somewhat inadequate 'Sprinter' DMUs. After being greatly reduced in numbers by withdrawals, the survivors were concentrated at Tinsley depot by September 1986, where ballast duties were expected to become their primary work. Two months later the remaining '45/1' examples arrived from Toton depot to be employed on similar duties after the InterCity, Provincial and Railfreight sectors had all failed to express any interest in using the class. Withdrawals continued at a rapid pace, but the remaining 'old ladies' of this fleet have continued to demonstrate their usefulness almost to the point of defiance — passenger operation was supposed to cease by May 1987, but the class still worked several regular turns during the summer, and the Class 45/1 survivors regularly worked out of St Pancras despite being officially banned south of Derby! During October 1987 Nos 45022/29/ 34/40/66 became Departmental Nos 97409-13, limited to 45mph and intended principally for use on ECML engineering duties. The class was expected to disappear by May 1988, with no regular diagrams planned, but the events of the previous year repeated themselves. Twenty examples (apart from the Departmentals) survived, and appeared on several NE/SW and St Pancras-Derby/Nottingham jobs due to a desperate lack of Sprinter units and other motive power shortages elsewhere on BR.

Disappearance of the newspaper traffic in July 1988 finally pushed the 'Peaks' out of main line service, and the survivors were switched off by 2 August. Fortunately the

Below:
Commencing in 1973, a total of 50 'Peaks' were converted to provide electrical train heating, using a separate Brush alternator and rectifiers, and the sub-class was known as Class 45/1 under the TOPS system. No 45104 (formerly D59) *The Royal Warwickshire Fusiliers* passes Luton with an up express on 7 April 1977. *K. Lane*

Left:
No 45127 arrives at Huddersfield with the 11.00 Scarborough-Newcastle on 21 September 1984. The ETH connections can be clearly seen, with a socket below the left buffer and a heavy plug and cable near the right-hand buffer. *S. R. Batty*

Below left:
The former D62 seen again, this time as No 45143 *5th Royal Inniskilling Dragoon Guards* hauling the 11.00 Scarborough-Bangor at Heaton Lodge, near Mirfield, on 26 February 1985. These duties were shared with Class 47. *S. R. Batty*

Charter Unit needed another 'celebrity' locomotive and No 45106 was chosen for these duties. As well as being an attractive source of motive power for various specials — especially after being repainted at Tinsley TMD — she also performed on regular BR duties, especially on the Midland Main Line, but timekeeping was not particularly good and No 45106 was relegated to less onerous duties. Fate took a hand on 3 February 1989 when the locomotive caught fire, leading to withdrawal on 20 February. No 45128 was then partially renovated, but returned to the scrap line when the cost of necessary repairs became prohibitive. Two Class 44 locomotives have been preserved (Nos 44004 and 44008), No 45060 (ex-D100) *Sherwood Forester* was purchased for restoration in early 1987 at Matlock and was joined in early 1989 by No 45135 *3rd Carabinier*. Attempts are now being made to secure funds for the purchase and restoration of Class 46 No 97404 (ex-46045/D182). Presently owned by the Derby Research Centre, and Nos 45108 and 45133 are also possible preservation candidates — both locomotives are presently stored at March.

Below left:
After being officially removed from main line passenger duties the 'Peaks' were regularly used on various freight duties. No 45012 passes Oakenshaw, near Wakefield, with 7M41 Dewsbury-Earles Sidings cement wagons on 4 June 1987. The locomotive looks to be in very good condition, despite its 26 years. *S. R. Batty*

Below:
Another view of the Dewsbury-Earles Sidings cement working, this time with Departmental No 97413 *Amethyst* in charge on 1 June 1988. Although limited to 45mph, this locomotive (the former 45066/D114) had plenty of power available to lift her train on to the York-Sheffield line at Moorthorpe. The unofficial names were painted on by Tinsley depot staff as the survivors were gathered at the shed, and 19 of the 22 survivors were similarly embellished by late 1987. *S. R. Batty*

The Production 'Deltics'

3 English Electric Co 'Deltic' 3,300hp Co-Co Type 5

Class: BR Type 5, later Class 55
Purpose: Express passenger locomotives
Introduced: 1961
Original numbers: D9000-D9021
Total built: 22

Details of Nos D9000-D9021
Engines: Two Napier Deltic 18
No of cylinders and cycle — 18 (opposed) two stroke
Max cont rated output — 1,650hp and 1,500hp each
Main generators: Two English Electric Type EE829
Traction motors: Six English Electric Type EE538
Type of suspension — nose
Type of gear drive — single reduction

Performance: Max tractive effort — 50,000lb at 22.3% adhesion at 2,400A main generator
Cont tractive effort — 30,500lb at 32.5mph at 1,650A main generator
Rail hp at cont rating — 2,640hp
Full engine output — available between 18.5 and 100mph
Braking: Air for loco, vacuum for train giving a brake force of 86% of loco weight in working order
Speed: 100mph (max permitted service)
Curve: 4 chains without gauge widening at dead slow speed
Train heating equipment: Spanner Mk II with a steaming capacity of 1,500lb/h
Tank capacities: Engine and boiler fuel 900gal
Boiler water — 640gal

Above:
A typical mid-1960s West Riding scene, with 'Deltic' No D9004 *Queen's Own Highlander* leaving Ardsley Tunnel with the 12.14 Leeds-King's Cross. The 'Deltic' has the original yellow warning panel, four windscreen wipers, regimental crest above the nameplate and repositioned BR badge below the cabside number. *E. Treacy*

The production 'Deltics', the Type 5 Co-Co thoroughbreds of the East Coast main line until the coming of the HSTs, reached the end of a brilliant working career in 1981. The high cost of maintenance for this small fleet of specialised locomotives could no longer be justified in view of BR's financial difficulties. When new, the locomotives were maintained (except for some mechanical portions) by engineers from EE Co and Napiers. This arrangement was retained under the initial contract for a period of five years and was subsequently renewed. It was geared to the *performance* of the locomotives, and if this fell below a specified mileage, a penalty clause in the contract reduced the sum BR paid to EE Co and Napiers for maintenance and engine overhauls.

The production 'Deltics' varied in both appearance and constructional details from

Below:
Production version of the English Electric Co 'Deltic' 3,300hp Co-Co, as in mid-1970s condition. Weights on the axles varied considerably during their lives, due to various modifications. *R. S. Carter*

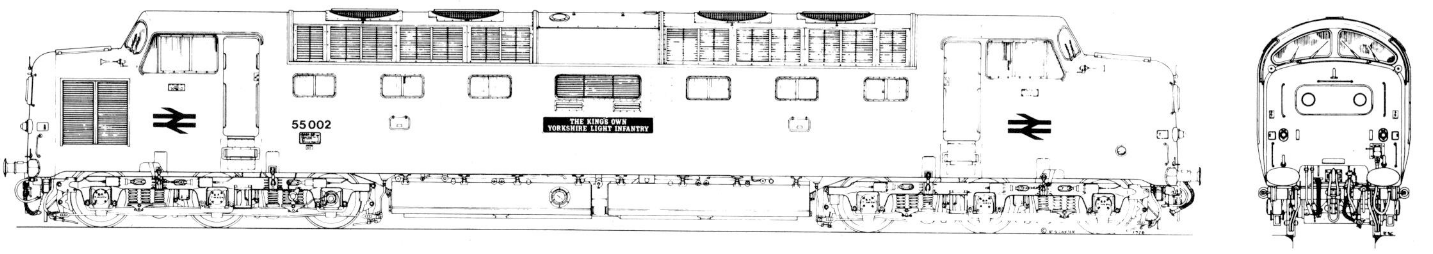

Left:
An unidentified member of the class races into Welwyn South Tunnel with a Newcastle-King's Cross express in April 1962. The locomotive is fitted with experimental roof-mounted air horns. *BR*

the privately-built prototype (see Section 4 of Book one of this series). The prototype, which had clocked-up some 400,000 miles in service on BR by the time the production contract was signed, featured a body profile which gave it a somewhat restricted route availability. The 22 slimmed-down production locomotives were some 7 tons lighter and 3ft 6in longer, with a bodyside contour that was swept-in below the waist to improve clearances. The typical EE Co nose end layout was retained, without the large headlight, but with a four-character headcode panel incorporated; no gangways were provided in the nose ends.

Production of the first few of the 22 locomotives was delayed because trouble was experienced with fractures on the bogies of some new EE Co Type 3 Co-Cos, then being introduced on the Great Eastern section. These bogies were of the same design as those on the 'Deltics' under construction. Thus the first locomotive arrived at Finsbury Park in March 1961; too late for the new fleet to have any influence upon the timings for that summer's timetable, much to the ER Board's regret.

Despite the hard work achieved by the prototype, there were still some teething troubles with the production locomotives and further modifications were necessary to the bogies (cast steel frames replaced fabricated parts) and to the Napier engines. The traction motors were prone to flashovers and many alterations were made to these as time progressed. Adverse factors were the very high noise level both to lineside dwellers and passengers at stations, and to the enginemen themselves. The noise in the cabs, a deep drone, was disturbing to the men, and for a period they were issued with earmuffs to combat this, pending better sound insulation. Fumes from the Napier engines were on occasions so prolific that the tunnels outside King's Cross station, for example, resembled the foggy air of mystery characteristic of former steam days! Another feature was their proneness to wheel-slip, sometimes at speed.

Having listed their weaknesses one must immediately remind oneself of their *strengths*. The 'Deltics' achieved high utilisation and they placed the East Coast route in the forefront of diesel-electric traction, in terms of speed and haulage, for 20 years. Renewed maintenance contracts kept the fleet in good shape, and the fitting of ETH in 1970 onwards gave Doncaster a chance to

Left:
The overall rail blue livery with full yellow noses was not as successful as the original two-tone green paintwork. Locomotive numbers were transferred to the bodysides behind the cab doors — initially at both ends — and the BR lion & wheel crest was replaced by the new 'double arrow' symbol. No D9000 *Royal Scots Grey* is seen in this condition hauling an up express through Sandy on 8 March 1970. *A. N. Pierce*

Above:
Sandy station again, with No 9005 *The Prince of Wales's Own Regiment of Yorkshire* passing through with the 14.00 King's Cross-Edinburgh on 11 January 1971. The Gateshead fleet was named after English regiments, the Haymarket locomotives were named after Scottish regiments and the Finsbury Park 'Deltics' after famous racehorses. *J. H. Cooper-Smith*

Left:
No 9001 *St Paddy* eases the 11.00 King's Cross-Newcastle into York on 28 October 1971. To keep up with developments in speed and comfort of rolling stock on the ECML during their working lives, the 'Deltic' fleet was fitted with dual braking (1967/68) and electric train heating equipment (1970/71).
J. H. Cooper-Smith

Above right:
Class 55 No 55011 *The Royal Northumberland Fusiliers* moves out of Newcastle station after arrival with the 13.25 stopping train from Edinburgh on 3 March 1981. The regimental crest, normally fitted above the nameplate, has been removed. *B. J. Beer*

Right:
The 'Deltics' ended their days on semi-fast trains such as the King's Cross-York stopping services, and No 55009 *Alycidon* is seen here accelerating away from Selby with the 10.45 York-King's Cross on 5 May 1981. Notice the ETH plug above the right-hand buffer, the blanked-off route indicator panel with marker lights, single windscreen wipers, repositioned air horns and white cab window surrounds — this latter was a special Finsbury Park treatment applied during the last years of 'Deltic' operation. *S. R. Batty*

bring them right up to scratch after nearly a decade of intensive main line service. Basically it was the bogies and the Napier engines that dictated the periods between shop visits. Air brakes were fitted for train braking in 1967/68, and sundry other modifications were made, including repositioning the air horns (see illustrations) and modifying the cabside windows.

The 'Deltics' created an enthusiast following for themselves like their Gresley 'A4' Pacific forebears, and the sights and sounds of these majestic machines racing along with apparent ease will long be remembered by a great many people. Locomotives Nos 55002/9/15/16/19/22 have been preserved in addition to the prototype.

Second Generation Types 4 and 5

Greater power and reduced weight were the essential requirements for the second generation of Type 4 diesel-electrics. The mixed traffic concept remained the ideal (only the 22 highly specialised 'Deltics' were express passenger locomotives) and BR quickly set about finding a suitable design, whilst the production order for the heavyweight D1 and D200 classes were fulfilled. Private industry, eager for new orders, collaborated and built speculative prototypes, whilst BR's own engineers and draughtsmen classified what was required in terms of engine, transmission and mechanical parts. At one stage, with the WR hydraulics already losing favour, and soon to be completely written off, a twin high speed engined diesel-electric version of the C-C 'Western' class was seriously considered for construction at Crewe; but large, single medium-speed diesel engines were considered more reliable and an outline specification was accordingly produced for a single-engined 2,700hp Co-Co. The design finally settled-upon was the Brush/Sulzer Type 4, D1500 class, and it is this locomotive which still dominates main line BR diesel traction today, with over 450 still in service at the time of writing. The present day Class 56 is essentially a freight locomotive, whilst its successor the Class 58 is officially described as a 'third generation' design for freight traffic. In 1985 a proposal for a Class 48 replacement design for the Brush machines was drafted, but this was effectively cancelled during the following year. At the present time a case is being drawn up for the building of 80 Class 41 machines of 2,200-2,500hp output, primarily for use with BR's coal traffic, and no plans for the replacement of Class 47 are being progressed. Meanwhile, the first heavy-haul Class 60 locomotives are under construction, and a further variant — Class 65 — is under consideration for use on higher-speed heavy-haulage duties.

Below:
The Brush/Sulzer Type 4 design typifies the second generation of BR diesel motive power design, with simple, rugged features and lack of cab nose-ends. Although the Sulzer engines have not been an unqualified success, the Class 47 fleet eventually totalled some 512. Many of these continue to give good service today in several different guises in all parts of the country. No 47097, with unusual small marker lights, stands at Westbury depot on 17 October 1979.
G. F. Gillham

4 Brush/Sulzer 2,750hp (2,580hp), 2,650hp Co-Co Type 4

Class: BR Type 4, later Class 47/48*†
Purpose: Mixed traffic locomotives
Introduced: 1962
Original numbers: D1100-D1111, D1500-D1999
Total built: 512

Details of Nos D1500-D1520* (first batch as built)
Engine: Sulzer 12LDA 28C
No of cylinders and cycle — 12 four-stroke
Max cont rated output — 2,750hp at 800rpm
Main generator: Brush TG 160-60
Traction motors: Six Brush TM 64-68
Type of suspension — nose
Type of gear drive — straight spur
Performance: Max tractive effort — 55,000lb at 21% adhesion at 3,390A main generator
Cont tractive effort — 30,000lb at 27mph at 2,130A main generator
Rail hp at cont rating — 2,080hp
Full engine output — available between 12.8 and 85mph
Braking: Air for loco, vacuum for train giving a brake force of 85% of loco weight in working order
Speed: 95mph (max permitted service)
Curve: 4 chains without gauge widening at dead slow speed
Train heating equipment: Steam — Spanner Mk III with a steaming capacity of 1,850lb/h
Electric — Brush TG-160-16 generator at cont output of 320kW at 800V
Tank capacities: Engine and boiler fuel — 810gal
Boiler water — 1,250gal

*Details are given of the initial batch of locomotives only. In this series there have

Bottom:
This drawing depicts the '47/0' sub-class; many later variants exist and the main differences are described in the text. These variations of course affect weights upon individual axles to some degree. *R. S. Carter*

Below:
The lack of any gangway doors allowed the use of a neater one-piece headcode panel on all the Brush/Sulzer Type 4 fleet right from the start of construction, but these were gradually taken out of use as the years passed. Class 47/4 No 47471 *Norman Tunna GC* passes Mossley with the 10.05 Liverpool-Newcastle on 3 March 1984. *S. R. Batty*

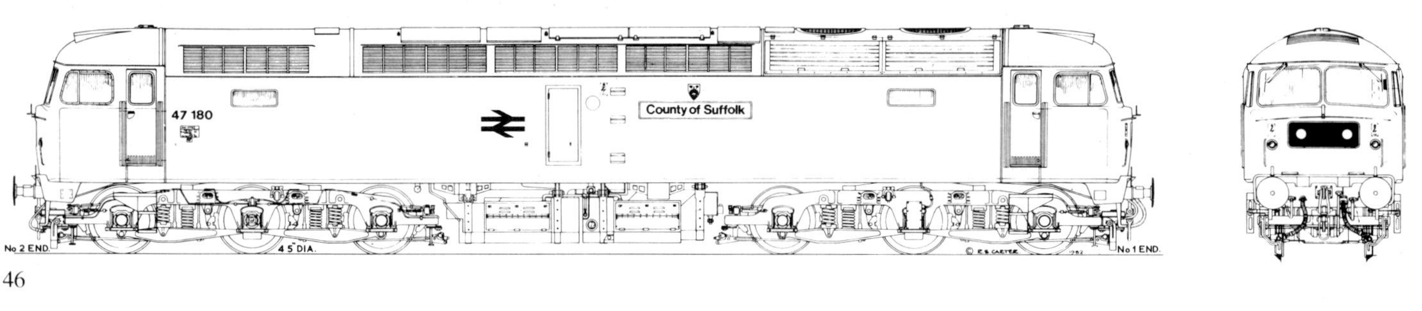

been numerous detail differences, and today's class fleet varies in so many respects that space prevents a description of anything other than a broad description in the text.
†Nos D1702-D1706 Vee-type engines.

Every so often in the on-going saga of British locomotive design a class appears upon the scene which is taken up as an acceptable standard, and built in considerable quantity. Examples in steam days were, of course, the GWR pannier tanks and 'Hall' class 4-6-0s and the LMSR Stanier 'Black Fives' and 8F 2-8-0s along with the MoS Riddles wartime 'Austerity' 2-8-0s. To the Brush/Sulzer Type 4 Co-Co diesel-electric must go the honours in recent times.

The background to the introduction of this versatile design, with an initial order for 20 locomotives (replacing the last 20 'Peak' class 1Co-Co1s) is described on page 37. It represented a much-desired improvement in terms of power/weight ratio and constructional techniques when compared to the two pilot scheme Type 4 diesel-electric designs. It was the ER in particular that had pressed the case for a Co-Co design with greater horsepower, when turning down the BTC offer of some new 'Peaks' made to them, in 1959. The ER had a good relationship with Brush, and this enterprising manufacturer took the initiative created by the ER, to commence the design of such a locomotive, using power equipment basically similar to the 'Peak' class locomotives then on its order book (see previous section). It was the successful uprating of the Sulzer 12LDA 28A 12-cylinder engine to 2,750hp (becoming 12LDA 28C) that encouraged Brush to pursue the design, as indeed it had also encouraged BR and the Birmingham Railway Carriage & Wagon Co and AEI, who were busily developing the *Lion* prototype; which alas did not completely come up to BR's needs. Following upon the BRCW declining an invitation to assist in the construction of the required fleet of 500 or more new standard Type 4s (the company was in the throes of financial collapse by then), BR approached Brush to produce an alternative to *Lion*, possibly also Sulzer-engined, but nearer to their requirements. In fact, Brush produced several alternative schemes, not all using the Sulzer engine, but the fact that some equipment was already on order from them for the final batch of 'Peaks', seems to have swayed the decision in favour of Sulzers.

The 'Peak' equipment was to be housed within a radically new and lightweight body, of monocoque, stressed-skin construction. Flat-fronted cabs were stipulated by BR and for the initial batch of 20, electrical train

Below:
Only the first 20 locomotives of the fleet were originally fitted with electric train heating equipment, the remainder carrying steam boilers or not having any heating equipment at all. A substantial conversion programme was then initiated to produce a sub-class capable of hauling BR's new fleet of electrically-heated rolling stock, the refurbished examples becoming Class 47/4. No 47569 hauls the 13.24 Llandudno-Scarborough around the curves near Mossley on 3 March 1984. *S. R. Batty*

47 066
Class 47
Weight tons 117
Brake force tons 60
RA 6
Max speed mph 95

Above left:
Class 47/0 No 47142 approaches Blea Moor Tunnel with a diverted newspaper van train on 11 December 1983. Many of the '47/0' sub-class carried steam-heating boilers, but this particular locomotive was not fitted with any heating equipment. *S. R. Batty*

Left:
An interesting close-up of No 1 end cab of No 47066, photographed on 1 August 1975 at Penzance. The previous fleet number — 1650 — can be seen through the rather battered paintwork on the cabside and also below the new TOPS number, and the details on the identification transfer below the number can also be seen. Also shown is the complicated design of brake rigging fitted to the Commonwealth bogies. *B. Morrison*

Above:
Blea Moor Tunnel again, with Class 47/4 No 47532 emerging on the diverted 09.59 Liverpool-Edinburgh on 6 May 1984. The large areas of blue paint made for a rather anonymous appearance, and the all-yellow cabs made the front end perhaps rather too plain. *S. R. Batty*

heating was provided in addition to the steam boilers. Under the agreement reached with Brush, BR had a licensing clause to build locomotives of the same type, at Crewe works, using Brush-designed and manufactured electrical equipment. A total of 512 was built by the two works.

With so large a class, and the passage of time, a great many variations now exist within the fleet. It would require a complete book of its own to attempt to describe these in any detail, and the reader in search of such information is referred to the list of books on page 21. The only large blot on the history of this useful class of locomotive concerned the Sulzer diesel engine and occurred during the years 1965-69. The full story is well described, with fullest technical detail, in Chapters 4 and 5 of *Class 47 Diesels* by Taylor, Thorley & Hill, and only a brief resumé is given here. Problems with the fabricated crankcases and internal joints led to a modification programme which involved much rewelding and rebalancing of the twin crankshafts and a reduction in maximum engine speed from 800rpm to 750rpm, and this latter feature led to a reduced engine maximum output of 2,580hp. BR, with hindsight, admit that the choice of the Sulzer engine was a mistake and it is a matter of some regret to them that this large family of locomotives do not have 2,700hp or plus EE Co engines. But I anticipate the story!

The main details of the class, as introduced are given at the head of this section. An important feature was the use of manganese cast steel Commonwealth bogies. This was acknowledged to be an extremely well-proven design, which whilst not being lightweight in concept, was much liked for its rugged form of construction and very tolerable ride quality. Despite this there were to be problems with fires and fractures, now largely cured.

Intensive testing of selected locomotives of the initial batch, including No D1500 on Dynamometer Car trials on the WR South Devon banks, revealed the soundness of the basic design; although there were the seemingly inevitable teething troubles. A regrettable decision made by BR was to cease fitting electrical train heating for further construction (as steam heating was envisaged as continuing for the immediate future), and this was to involve BR in a lengthy conversion programme (with some permanent loss of main engine output) in later years.

Quantity production by the two works meant that the class soon became a familiar sight in many parts of Britain. Modifications were made to the traction motors and it became necessary to fit air brakes for freight train work and for the new Mk 2 rolling stock. A significant development was the introduction in collaboration with the CEGB, of the 'merry-go-round' coal train for feeding fuel to the power stations. These mgr trains were required to move steadily at extremely low speed whilst loading and unloading took place. Brush developed a slow speed control which allows the locomotives so fitted to move at 0.5mph; this was developed to be fully automatic.

Five locomotives were chosen by BR for fitting with the French-built Sulzer Vee-type engine, with 12 cylinders. This was the 12LVA24 engine and Brush fitted it to Nos D1702-D1706 (later these became Class 48). The engine produced 2,650hp, ie 100hp less than the twin-bank engines. The five locomotives ran in this form until 1969 when they were converted to Class 47s and their engines are now running in some SNCF

Above:
In 1974 the badly-damaged No 47046 was rebuilt as a test-bed for the GEC/Ruston Paxman 16RK3CT engine which was destined for use in the Class 56 fleet, and the modified locomotive was then renumbered to 47601. Seen at Tinsley depot on 26 September 1976, this view gives an impression of the length of these locomotives, and also shows the modified roof panels and fuel tanks fitted during the conversion. Further work has seen this locomotive fitted with the 12RK3CT engine (as in the Class 58 fleet) and renumbered again to 47901. *P. D. Hawkins*

Centre left:
The sub-class 47/3 is essentially a freight locomotive, having no train-heating equipment and carrying rather smaller fuel tanks. All are equipped for slow-speed control, but many of their duties have been taken over by Classes 56 and 58. Class 47/3 No 47306 heads westwards through Gilberdyke, between Hull and Doncaster, with a short freight on 13 July 1984. *S. R. Batty*

Left:
The Brush/Sulzer Type 4s perform passenger duties alongside their electric brethren on the West Coast main line, regularly operating InterCity traffic away from the electrified areas. Class 47/4 No 47483 moves away from the 11.05 Ayr-Euston at Preston on 20 September 1985. *S. R. Batty*

diesels in France where numerous locomotives have these Vee-type engines.

It is significant that EE Co engines have been used in recent BR construction, and that two experiments have been applied, with re-engining to one Class 47. This is currently No 47901 of Class 47/9 and it is fitted with a 3,250hp Ruston-Paxman 12RK3CT 12-cylinder engine which is very similar to the unit fitted to Class 58 locomotives. Effectively a

Above:
On a similar duty, Class 47/4 No 47476 arrives at Poulton-le-Fylde with the 14.16 Blackpool (North)-Euston on 27 August 1986. The blanked-off headcode panel and recently fitted headlight are clearly visible. *S. R. Batty*

Centre right:
One of the early alternative liveries produced at Stratford depot, as applied to Class 47/4 No 47583 *County of Hertfordshire* in July 1981. Apart from the two horizontal white bodyside lines and the style of numerals, this scheme is substantially the same as today's livery for much of the Class 47 fleet. At the time of the photograph No 47583 has the headcode panels blanked off and marker lights fitted. *J. E. Augustson*

Below right:
InterCity livery is very colourful and enhances the appearance of the '47s' greatly. The roof and upper bodysides are slate blue and the lower bodyside is off-white with a broad red band. The cab roofs are yellow, the window area slate blue and the lower half also yellow. Class 47/4 No 47515 *Night Mail* awaits departure from Derby with the 22.20 Newcastle-St Pancras postal on 25 September 1986, the day of the locomotive's naming. *C. J. Tuffs*

Type 5, this locomotive was built as No D1628, which as No 47046 was severely damaged in an accident at Peterborough in September 1974. Crewe Works rebuilt the locomotive as Class 47/6 No 47601 with an EE/GEC 16CSVT 3,250hp engine to assist in development work on the planned Class 56 power unit. The present engine was fitted (and the locomotive renumbered) for work on the Class 58 project. At present No 47901 is being used to evaluate various types of piston design for this class.

ScotRail
47 637

47 660

Left:
The ScotRail Class 47 livery is similar to the InterCity version. Class 47/4 No 47637 approaches Edinburgh Waverley station from Haymarket shed on 14 December 1986. (This locomotive also carries the Highland Rail 'Stag' logo on the bodyside.) The original InterCity and ScotRail liveries both have quite large areas of the cab fronts painted in a dark colour instead of the usual all-over yellow, and, as this photograph shows, the original yellow paintwork on the cab roof has been obliterated by dirt, leaving only the lower part of the cab front in the high-visibility yellow. The headlight, however, is very powerful and gives plenty of warning of the locomotive's approach. *W. A. Sharman*

Below left:
The last six Class 47/4 conversions were originally authorised for use on the Trans-Pennine (Northern) route from Newcastle to Liverpool via Leeds and Huddersfield, and it was originally intended that these would be painted in a suitable livery to match the Trans-Pennine coaching stock. Unfortunately this was not done, and all six emerged from Crewe works in standard 'large logo' livery. The first of these, No 47660 (ex-47155), passes Sandal, Wakefield, with the 09.43 Newcastle-Penzance on 31 January 1987. In the main, these locomotives have seen use on InterCity rather than Trans-Pennine duties. One locomotive, No 47475, was repainted in 'Trans-Pennine' livery in early 1989.
S. R. Batty

Top right:
Crewe depot's No 47503 gets away from Shipley with the 10.42 Leeds-Carlisle on 5 August 1987. Today's standard livery suits the Class 47s quite well, giving the cab fronts a particularly neat and pleasing appearance.
S. R. Batty

Right:
A total of 16 Class 47s were modified for push-pull working between Edinburgh and Glasgow, forming sub-Class 47/7. This view shows No 47701 (later named *Saint Andrew*) shortly after conversion in 1979, and illustrates well the new control jumper cables, the original headlight fitting and the ETH equipment. The '47/7s' were converted from Class 47/4 locomotives. *B. J. Nicolle*

Below right:
Eight years later, and the '47/7s' have been considerably smartened up. Nos 47702 *Saint Cuthbert* and 47712 *Lady Diana Spencer* stand outside Haymarket depot on 18 April 1987, bearing full ScotRail livery and later-pattern headlamps. No 47712's headcode panel has been blanked off, but that on No 47702 has been completely removed to make a flush cab front. *S. Widdowson*

One more modification to a Class 47 deserves mention. This was No D1938, fitted new with push-pull equipment and jumper cables, in connection with trials for a proposed Paddington-Birmingham high speed push-pull service. Nothing came of this, but in 1981 a total of 12 locomotives were fitted for push-pull working between Glasgow and Edinburgh and were reclassified as '47/7'. Four more conversions were carried out during 1984/85 and a further three were considered for working between Leeds and Harrogate after the King's Cross-Leeds line was electrified, but this idea was dropped.

As already mentioned there are innumerable detail differences amongst members of this large class, but for convenience the basic variations are sub-classified by BR, as follows:

Class 47/0: Steam heating (isolated). Some with no boilers.
Class 47/3: No train heating. Used basically for freight duties and some fitted for mgr work.
Class 47/4: Electric or dual train heating.
Class 47/7: Push-pull fitted version of the Class 47/4.
Class 47/8: Class 47/4 locomotives with long-range fuel tanks of 1,295gal capacity.
Class 47/9: Re-engined (see earlier).

The Class 47/4 conversion programme was completed in early 1987 when No 47665 emerged from Crewe Works, almost exactly one year after three of the 'generator' ETH locomotives, Nos 47405/14/16, had become the first members of the class to be withdrawn. Locomotives Nos 47401-20 were the original members of the fleet, then numbered D1500-19, and their high mileages and DC generator-driven ETH system (as opposed to the alternator-powered systems of later examples) apparently conspired to make for early withdrawal. A dire shortage of spares amongst the class generally led to more withdrawals from 1987 onwards, and by February 1989 only eight of this batch of the first 20 locomotives was left in service. The year 1988/89 was expected to see 20 withdrawals just to provide spare parts for the rest of the fleet, and the displacement of the class from passenger duties makes this an extremely likely happening. Since late 1987 cash limits have been enforced on repairs, the remaining '47/0' and '47/3' locomotives have been limited to 75mph and 13 Class 47/0 have been restrained to only 60mph. Further overhauls to '47/0' locomotives have been provisionally arranged for 1989/90, but this extra expenditure is very vulnerable to cutting-back if any Sector tightens its purse strings. On the brighter side, many of the class have been painted in assorted InterCity, ScotRail and Railfreight liveries which have enhanced the appearance of these already attractive locomotives. One locomotive — No 47475 — has been given Trans-Pennine livery, as was originally to have been applied to Nos 47660-65. Names have been given to many of the class, and in many cases the nameplates have been removed and transferred to other locomotives upon conversion or reallocation to another business sector. Collision damage has led to the recent withdrawal of the first Class 47/3 locomotive, and overhauls on three further members of the sub-class have been cancelled. The '47/7' fleet remains intact — fire-damaged No 47713 was replaced by converting No 47497 to No 47717 in 1988 — but their Scottish duties are being taken over by Class 158 'Sprinters' and the locomotives are being transferred south originally for Waterloo-Exeter duties during 1989. The Class 47 locomotives do not have a good timekeeping record on this route, and they may well be used on other services to better effect.

Many of the class are now working from Immingham depot, and during 1988 a proposal was made that some of the class — perhaps 10 examples — should be converted to a high-adhesion specification for working the heavy oil trains found in the area. Long-range fuel tanks of 1,295gal capacity would further increase their usefulness, but the plan was altered and in early 1989 the first members of sub-class '47/8' appeared, and were allocated to the InterCity sector as long-range locomotives. The first 20 are merely renumbered Class 47/4 machines which already had the 1,295gal twin fuel tanks fitted, but a further 24 are to be converted at appropriate examinations. The Brush-Sulzer Type 4 will be around for a good few years yet, giving their versatile performances across the entire BR system. But, as stated earlier, one cannot help wondering how much the class would have benefited from having an English Electric engine fitted right from the start of production, instead of using the aged Sulzer product which needed extensive modification and derating.

Top:
The 'Generators' (Nos 47401-420) have now been officially retired from frontline passenger service, but No 47406 *Rail Riders* made a fine sight on 17 February 1988 heading past Mirfield with the 12.25 Newcastle-Liverpool Trans-Pennine duty.
S. R. Batty

Above:
Another 'Generator', No 47407 *Aycliffe*, passes Bradley Junction with the 12.25 Newcastle-Liverpool on 30 September 1987. A full set of Trans-Pennine-liveried rolling stock is a fine sight, and a similarly-painted batch of locomotives to operate these services would have created a very pleasing effect, but no doubt the needs of locomotive utilisation would have made this difficult to achieve at all times. *S. R. Batty*

Right:
An unusual one-off at a time when BR is attempting to standardise its liveries is No 47475 in Provincial Sector Trans-Pennine colours. The locomotive is seen here at Marsden with the 09.01 Liverpool-Newcastle on 21 April 1989. *S. R. Batty*

5 English Electric Co 2,700hp Co-Co Type 4

Class: BR Type 4, later Class 50
Purpose: Mixed traffic locomotives
Introduced: 1967
Original numbers: D400-D449
Total built: 50

Details of Nos D400-D449
Engine: English Electric 16C5VT
No of cylinders and cycle — 16 four-stroke
Max cont rated output — 2,700hp at 850rpm
Main generator: English Electric Type EE 840/4B
Traction motors: Six English Electric Type EE 538/5A
Type of suspension — nose
Type of gear drive — single reduction
Performance: Max tractive effort — 48,500lb at 18.8% at 2,400A main generator
Cont tractive effort — 33,000lb at 23.5mph at 1,800A main generator
Rail hp at cont rating — 2,070hp
Full engine output — available between 15 and 87mph
Braking: Air for loco, air or vacuum for train giving a brake force of 84.9% of loco weight in full working order
Speed: 100mph (max permitted service)
Curve: 4 chains without gauge widening at dead slow speed
3½ chains with ¾in gauge widening at dead slow speed
Train heating equipment: Electric, 320kW at 850V
Tank capacity: Engine fuel — 1,000gal

The production development of the privately sponsored EE Co Type 4 Co-Co prototype No DP2, these 2,700hp machines were required as a stopgap for the LMR/ScR West Coast main line between Weaver Junction and Glasgow; pending completion of the 25kV electrification. The English Electric Co produced the design in close collaboration with BR, and when compared to the prototype a number of differences resulted, making these locomotives initially more complex and sophisticated than No DP2.

In particular these machines turned out to be heavier and more expensive to build. They featured provision for weight transfer compensation on the cast steel bogies (of similar design to those of the 'Deltics' in other respects) and they had a flat-fronted cab superstructure — the first main line diesels built by EE Co for BR not to have nose ends. The cab design proposed by the EE Co was

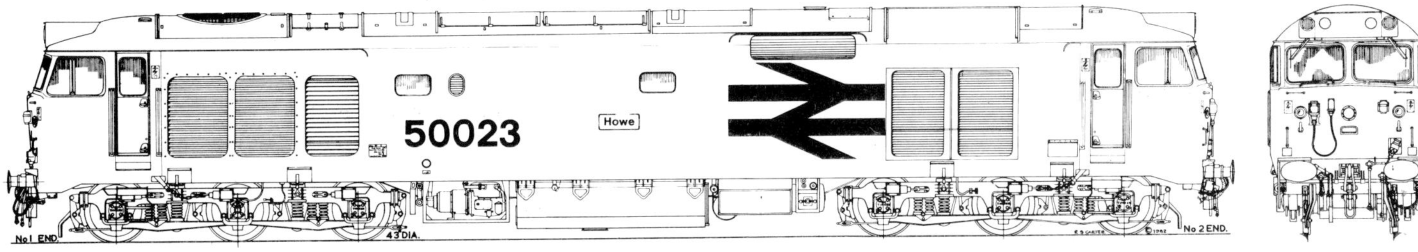

Below:
English Electric Co 2,700hp Co-Co Type 4, depicted in refurbished guise. *R. S. Carter*

Bottom:
The EE Co Type 4 was produced essentially as a stop-gap measure to improve services on the West Coast main line pending full electrification to Glasgow, and the class was regularly operated in pairs to produce the necessary horsepower to achieve the desired level of service. Nos 413 and 400 pass Warrington with the 14.05 Euston-Glasgow on 30 March 1971, a typical WCML service of the time. The class appeared new in the 'Rail Blue' livery with full yellow ends and 'double arrow' symbol, but these two examples have lost the 'D' prefix to their numbers.
J. H. Cooper-Smith

1V90

ay out
trains

Left:
After completion of the WCML electrification the 'D400' series (which became Class 50 in TOPS parlance) were dispersed to the Western Region. Seen on an early example of their new workings, No 50025 approaches Barnt Green with the 13.20 Liverpool-Plymouth on 6 June 1974, having worked to Birmingham on the same day's 07.40 Penzance-Liverpool. *P. D. Hawkins*

Below left:
Class 50 No 50030 *Repulse* arrives at Liskeard with the 13.59 Penzance-Bristol on 18 July 1978. The Western Region bestowed names on the class, many of which came from the diesel-hydraulic fleet which the Class 50s had helped to eliminate. The headcode panels had a relatively short life, officially being taken out of use by January 1976, and were soon replaced by marker lights; ultimately they were blanked off completely. *B. Morrison*

Above:
Nos 50045 *Achilles* (with very shabby paintwork) and 50040 *Leviathan* speed down from Whiteball Summit with the 08.30 Paddington-Plymouth on 8 April 1980. The front-end appearance would have been greatly improved if the multiple-control jumper cables had been placed at bufferbeam level instead of directly on the cab front. *D. Clow*

very attractive with twin wrapped-round Triplex windscreens, a four-panel route indicator below the windows and a neat arrangement of mu jumper cables. Unfortunately BR had produced its own mock-up for a future cab design, in collaboration with interested parties such as the Safety and Accident Prevention people at Derby. This mock-up was disliked for its external appearance by the Design Panel, but Derby won the day, with the result now carried on the Class 50s (as they are now known), complete with the rather ugly roof-mounted indicator box and air horns.

The 2,700hp 16-cylinder turbo-charged and intercooled 16CSVT diesel engine was basically similar to the one in the prototype, as was the main electrical machinery. An electrical generator was provided for train heating, and the provision was made for multiple-unit working of up to three locomotives, for push-pull working and for slow speed control for 'merry-go-round' coal trains. The traction motors were basically similar to the EE Co Type 3 Co-Cos, D6700 class (now Class 37).

The sophisticated items of advanced technology applied included dynamic braking; automatic control of tractive effort and wheelslip detection, inertia filtration, and solid state devices in the power control equipment. Thus these 50 locomotives had a number of features which would require extra maintenance and attention.

The D400 class began to appear from Vulcan Foundry in late 1967, and they sported the rail blue livery with yellow ends when new. The first two, Nos D400/1 had the

Below:
Class 50s under overhaul at BREL Doncaster during September 1982, with a complete power unit on view. *S. R. Batty*

Left:
Although fitted with some very sophisticated equipment, especially control gear, many of these items caused far more problems than was considered justifiable. A refurbishment programme to improve reliability commenced during 1979, and took until 1983 to complete. Refurbished No 50047 *Swiftsure* emerges from Middle Hill Tunnel, Box, with the 09.10 Bristol-Paddington on 9 August 1980. All the class received headlights in connection with the rebuilding work and the majority emerged from Doncaster with the new standard 'large logo' livery, but six early examples (including the one shown here) were initially repainted in the original standard scheme. *J. Chalcraft*

ugly mu jumper cables and sockets on the nose ends but subsequent locomotives emerged without them (although wired-up internally), and these were fitted later at Crewe, when it was decided to work the class in pairs on Anglo-Scottish duties. The D400s had a good high-speed performance, and drivers soon discovered their ability to slog along at 100mph or more, but their overall availability was revealed to be poor and there were soon troubles with their engines; requiring new cylinder heads to be fitted by Vulcan and Crewe. Nevertheless the class gave a good account of themselves prior to completion of the electrification of the West Coast main line.

The transfer of all 50 locomotives to the WR, once made redundant by electrification, was a planned intention when they were ordered. This meant that the 'Western' class diesel-hydraulics could then be withdrawn from the principal main line passenger services. This move began in late 1972 with Nos D400/1 going first for driver-training, to Bristol. The class were not in very good shape when received by the WR and failures on the road earned them a poor reputation, initially. In particular the prolonged 100mph running possible on the London end of the WR main line to Bristol revealed weaknesses in the traction motors, which had to be modified. A programme of refurbishment to increase reliability and simplicity was undertaken by Doncaster Works, and this was completed during 1983 when No 50014 was outshopped. At the same time the whole class was temporarily limited to a maximum speed of 95mph. By this time the entire class had been given names by the Western Region of BR which were reminiscent of the diesel-hydraulic 'Warship' class.

Locomotive No 50041 *Bulwark* was very badly damaged in a derailment outside Paddington station in November 1983, but she was repaired and returned to service by Doncaster Works. Alas, more recent years have not been so kind to the class and several examples have been withdrawn. Nos 50011 *Centurion*, 50006 *Neptune* and 50014 *Warspite* were taken out of service during 1987 due to poor mechanical condition, a lack of spare parts for other members of the class and a greatly reduced workload. Frame cracking has always been a problem (six of the class were stopped at Doncaster for attention to this fault in the summer of 1984), and No 50011 succumbed to this failure, and a worn-out engine, during February 1987. Five locomotives were withdrawn in 1988, and one has succumbed in 1989 at the time of writing. Doubts were cast on the class's future in 1986 when they were removed from Western Region InterCity duties, and when it was also expected that they would no longer be required on Waterloo-Exeter duties beyond May 1987. (This latter expectation thankfully proved to be hopelessly inaccurate.) Unfortunately, the Railfreight Sector shied away from taking any of the class due to their low tractive effort, high maintenance costs and poor availability — no less than 19 were out of service during mid-1986. Nevertheless, someone with high powers of persuasion managed to arrange for No 50049 to be modified to more of a freight locomotive specification during 1987, using some components from withdrawn No 50006. Class 50/1 No 50149 emerged later in the year for test operation, but the experiment was quickly deemed to be unsuccessful and was abandoned. Had the modifications been successful the class could have been used for hauling heavy stone and gypsum traffic, but the solitary No 50149 was quickly despatched out of the way to worry the china clay trains in deepest Cornwall. This move turned out to be of rather more use than was perhaps expected at first, as the locomotive was called upon to cover some duties left vacant after

Left:
Despite being purely Western Region locomotives from 1976 onwards, the class frequently worked north of Birmingham. The 14.00 Leeds-Brighton is clattered over Hare Park Junction, south of Wakefield, by No 50029 *Renown* on 26 January 1985.
S. R. Batty

Above:
In the writer's view the revised livery with black cab window surrounds and yellow area extended back to the rear of the cab doors is a great improvement over the original scheme. No 50020 *Revenge* hurries the 11.05 Paignton-Paddington along near Hele & Bradninch on 3 July 1985. A headlight has been fitted and the headcode panel has been blanked off, with marker lights fitted.
S. R. Batty

Right:
A very smart No 50010 *Monarch* speeds through Burbage Wharf with the 14.27 Paddington-Penzance on 7 September 1985.
B. J. Nicolle

two Class 37s were badly damaged in November 1987. No 50149 was re-converted back to standard specification in early 1989 and returned to service on Exeter line duties in February. Despite maintenance problems and variable availability, these locomotives are better timekeepers than the less powerful

Right:
The Network SouthEast livery has been applied to a number of Class 50s, and in the writer's opinion the visual effect is little short of disastrous. Most of the body is painted blue, with a broad white stripe along the length containing a narrow red stripe within. Originally these stripes were swept upwards on the cabsides, but the present version has them carried along horizontally to meet a small yellow area extended back around the cab side. No 50005 *Collingwood* sports the newer version on 5 April 1988, hauling the 11.05 Exeter St Davids-Portsmouth Harbour past Millbrook, near Southampton. *D. Brown*

Left:
The former Southern Region Waterloo-Exeter route has been a stronghold of the class for several years. No 50022 *Anson* thunders through the remains of Seaton Junction with the 16.17 Exeter-Waterloo on 21 June 1988. *S. R. Batty*

Class 47 machines which were expected to take over some of these duties in 1989.

During 1987 the class maintained its presence on the Exeter services by courtesy of the Network SouthEast Sector, but the InterCity and Railfreight businesses still had no work for the class. By April 1988 Laira depot at Plymouth had stopped performing 'F' examinations on the class, and a pessimistic estimate suggested that one example per month might be withdrawn during the rest of the year. This did not take place, and the class is now being used on duties for the Departmental, Parcels and Network SouthEast Sectors. The class probably does not have a long-term future, and withdrawals will be dictated by component failure, availability of alternative motive power and the workload of the owning Sectors.

Below left:
The same working on the following day, with NSE-liveried No 50003 *Temeraire* approaching Seaton Junction. The new livery gives the locomotive a very heavy appearance and is certainly not an improvement over previous schemes. *S. R. Batty*

Below:
The one-time freight version of the class, the former No 50049 now renumbered to 50149 *Defiance* and reclassified to 50/1. Seen at Exeter whilst awaiting trials on 14 November 1987, the locomotive boasts full new Railfreight livery with sub-Sector markings, small abbreviated cab-front number, repainted nameplates and cast-metal depot badge on the cabside. This locomotive has since been converted back to standard. *C. J. Marsden*

6 British Railways/Brush 3,250hp Class 56

Above:
Empty MGR wagons from Fiddlers Ferry power station pass below Warrington (Bank Quay) station on the first stage of their journey back to Yorkshire behind Doncaster-built No 56093 on 22 September 1981, soon after this locomotive entered service. It features large logo livery, full yellow cabs with black window surrounds, and silver roof. *P. Spilsbury*

Class: BR Class 56
Purpose: Heavy freight locomotives
Introduced: 1976
Original numbers: 56001 onwards
Total built: 135

Details of Nos 56001 onwards
Engine: Ruston 16RK3CT
No of cylinders and cycle — 16 four-stroke
Max cont rated output — 2,460kW at 900rpm
Main alternator: Brush BA1101A
Traction motors: Six Brush TM 73-62
Type of suspension — nose
Type of gear drive — single reduction
Performance: Max tractive effort — 270,000N (61,000lb) at 22% at 3,050A traction current
Cont tractive effort — 240,000N (54,000lb) at 27km/h at 2,730A traction current
Rail hp at cont rating — 1,790kW
Full engine output — available between 26-130km/h
Braking: Straight air and auto air for loco, auto air for train giving a brake force of 71.7% of loco weight in full working order
Speed: 130km/h (max permitted service)
Curves: 80m without gauge widening at dead slow speed
74m with 20mm gauge widening at dead slow speed
Tank capacity: 5,228 litres

BR's policy of building powerful diesel-electrics specifically for heavy freight duties resulted in the production, at a very slow rate, of Class 56 3,250hp Co-Cos, from 1976 to 1984.

The 1973 Middle East conflict and the subsequent rise to power of the oil sheikhs resulted in an urgent review of Britain's coal resources. Coal suddenly became a priority fuel once again and BR was faced with a need to strengthen its fleet to provide haulage of 'merry-go-round' coal trains to power stations. The Class 56 resulted from this need, and they were produced in something of a hurry; being regarded as an interim design. It is, however, as *Modern Railways* once described it 'a thoroughbred with an impeccable pedigree', being based upon the Brush Class 47 for its superstructure and with the long-developed EE Co medium speed

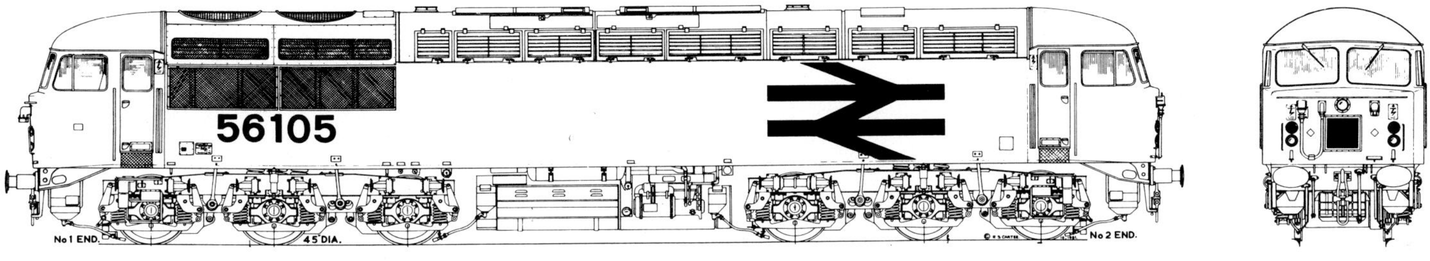

Below:
General arrangement of the BR/Brush Class 56 heavy freight 3,250hp Co-Co, showing one of the later variants. An enlarged front air horn grille was fitted from No 56056 onwards and is also carried by earlier locomotives that have suffered accident damage. *R. S. Carter*

WHARTON CRANES
SWL MAIN 45 TONNES

Above left:
The original livery of Rail blue with small symbols and numerals produced a very dull-looking locomotive. Nos 56038 *Western Mail* and 56037 *Richard Trevithick* haul a demonstration Port Talbot-Llanwern iron ore train through Fairwater, Cardiff, after the naming of the two locomotives on 5 August 1981. Note the smaller pattern of air-horn grille, round buffers and smaller type of headlight. *K. Hayward*

Left:
Locomotives Nos 56111-114 under construction at Doncaster during September 1982, all with oval buffers and the larger air-horn grille fitted from new to Nos 56056 onwards. *S. R. Batty*

Above:
During the 1984/85 miners' strike the class was used on several alternative duties whilst the MGR programme was severely curtailed. Pairs of locomotives regularly operated the heavy Teesside-Corby steel trains, and Nos 56007 and 56128 are seen at Brotherton with the returning empties on 25 October 1984. The Romanian-built examples all have heavy rubber surrounds to the windscreens and a different pattern of small cab-side window, as displayed by No 56007. *S. R. Batty*

engine. Slow speed control is fitted for mgr working.

The biggest surprise concerning the Class 56 was the decision to build the first 30 locomotives in Romania! This was due to a lack of manufacturing capacity at Brush Electrical Machines, who were the locomotive's designers, in collaboration with BR. A fast delivery was required and Romania could apparently promise this by building them at the Craiova works of Electroputere, using UK-supplied power equipment and some British components. The bogies are of Swiss design, except on one Doncaster-built locomotive, No 56042, which is experimentally fitted with prototype bogies of Derby design, known as Type CP1. The Swiss design was chosen because it was extensively used in Romania and was readily available.

Unhappily, the Romanian-built locomotives developed so many teething troubles that the desired early delivery did not come about and BR did not accept delivery of the first units until March 1977 (they had been on British soil since the previous August). This is not the time or the place to describe these problems and it must suffice to say that standards of workmanship and manufacture were decidedly poor on the first locomotives.

Doncaster works took over construction of the class after the initial batch, and the final 20 examples were built at Crewe. The Doncaster build was completed with No 56115 in early 1983, and No 56135 emerged from BREL Crewe in November 1984. Initially the class was divided between the depots of Gateshead, Healey Mills, Tinsley, Toton, Bristol and Cardiff for working heavy Merry-go-Round coal trains and also steel and stone traffic. By late 1987 the entire class was being operated from either Toton or Cardiff, with batches being outstationed to local depots (eg Blyth, Knottingley, Healey Mills, Bristol) nearer to where their traffic is operated. Approximately 20 of the class now carry names, and these are directly related to the type of traffic which is regularly hauled or to the area within which the locomotive is usually to be found.

The '56s' have performed some prodigious feats of haulage in connection with the steel and stone traffic, a typical example being the trains from Merehead quarry which started running in September 1983. These 4,500-ton gross trains contained a payload of 3,300 tons within 43 box wagons and required two locomotives for haulage. Events have moved on over the past five years, and the introduction of the Foster Yeoman Class 59 locomotives has taken much of this particular traffic away from the class and also pointed a finger in the direction in which future designs of heavy haulage freight locomotives must lie. Nevertheless, the Class 56 locomotives have proved themselves to be capable of handling all BR's heavy freight jobs, particularly the intensive MGR diagrams in Yorkshire, and the heavy oil trains found in South Wales and Immingham. The length and clean front-end appearance makes the class visually very attractive (especially when presented in the latest Railfreight livery) and the sight and sound of one of these locomotives working hard with a heavy train leaves a strong impression on a lineside observer. When working at high outputs, the hard, distinctive throb of the engine betrays the clear English Electric ancestry of the power unit, and there is every reason to believe that the class will enjoy many years of work yet. The proposed Class 41 locomotives are intended for the coal business and would be capable of hauling equal loads to the Class 56 but at reduced speeds, due to the reduced engine output of 2,200-2,500hp. Consequently the '56s' could be released for hauling faster traffic than is their lot at present, such as the newly-acquired ex-Freightliner traffic of the Speedlink service.

Right:
Crewe-built No 56118 bowls along the Castleford line at Milford Junction with a train of coal from the northeast for Ferrybridge power station on 18 March 1985. These locomotives were specifically designed for heavy-duty jobs and have enough power to accelerate these 1,400-ton trains rapidly up to 45-50mph. *S. R. Batty*

Below:
Class 56 No 56125 passes through the remains of Glasshoughton colliery with 7K83 Bowers Row-Drax on 29 October 1985. The fitting of round or oval-pattern buffers appears to be at random. *S. R. Batty*

Far right:
The driver of No 56112 opens the controller wide as he passes Royston Junction with a well-filled train of coal from Royston colliery for Drax power station on 21 August 1986. When under heavy load, the hard, solid throb of the GEC/Ruston engine leaves the listener in no doubt as to its English Electric origins which reach back to the late 1940s.
S. R. Batty

Below right:
The first of the Crewe-built batch, No 56116, eases a train of empties back into Ferrybridge 'A' sidings on 24 November 1986. The larger type of headlight fitted to most of the class is visible. *S. R. Batty*

56112

56116
116

Left:
A modified version of Railfreight livery used the same colours (grey roof and bodysides with full yellow cabs and black window areas) as before, but with a red band around the body at solebar level, and smaller 'double arrow' symbol. No 56089 shows these features as she approaches Eggborough power station with a loaded train on 4 June 1987. *S. R. Batty*

Below:
A selection of front ends at Knottingley TMD on 15 May 1988. The subtle variations in cab window treatment and air-horn grilles are apparent, and the three locomotives fitted with cab-to-shore radio aerials also have the metal buffer cowling which was subsequently abandoned. *S. R.Batty*

Right:
Not all coal is carried by MGR train! No 56132 *Fina Energy* crosses to the Sheffield line at Milford Junction with a train of coal containers on 24 May 1988. *S. R. Batty*

Below right:
In immaculate ex-works condition, No 56023 passes through Moorthorpe station with MGR empties on 11 June 1988. The Railfreight Sector livery consists of black and yellow cabs with a two-tone grey bodywork having the upper colour extended over the roof. A far cry from Rail blue! *S. R. Batty*

56132
132
20

Third Generation Types 4 and 5

Below:
The premier member of Class 58, No 58001, stands outside BREL Doncaster works soon after completion in early 1983. The modular assembly is apparent in this view, as is the strikingly different cab design. *BR*

7 British Railways 3,300hp Co-Co Class 58

Class: BR Class 58
Purpose: Heavy freight locomotives
Introduced: 1982
Original numbers: 58001 onwards
Total built: 50

Details of Nos 58001 onwards
Engine: Ruston 12RK3CT
No of cylinders and cycle — 12 four-stroke
Max cont rated output — 2,610kW at 1,000rpm
Main alternator: Brush
Traction motors: Six Brush TM73.62
Type of suspension — Timken suspension tube
Type of gear drive — single reduction
Performance: Max tractive effort — 267,000N (60,700lb)
Braking: Straight air and auto air for loco
Speed: 139km/h (80mph) max permitted service

In 1977 BREL requested the Locomotive Design Office at Derby to prepare a report and produce a feasibility study on a new low-cost Co-Co locomotive with export potential. The design suggested was based on a single underframe formed of rolled steel joists, with a non-load-bearing superstructure. This arrangement gave substantial savings when compared to the Class 56 design, being some 13-14% cheaper.

The main objectives sought after in the design were: economy of construction, minimum overall maintenance requirement, adaptable to export requirements with minimum modifications.

The steel underframe is designed to carry all static loads imposed by the locomotive equipment, and those arising from dynamic forces when the locomotive is in operation. The underframe is designed to take an end

Below:
General arrangement of BR Class 58 3,300hp Co-Co as built. *R. S. Carter*

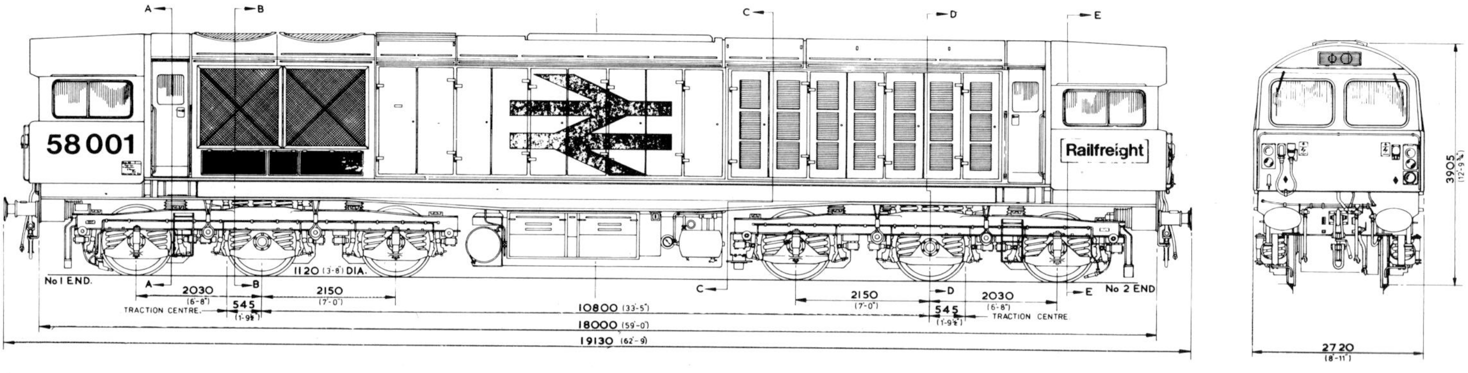

load of 200 tonnes at buffer height.

Each locomotive is made up of six modular units bolted to the rigid frame. These are interchangeable with a stock item so that collision damaged or failed locomotives will be out of traffic for a minimum amount of time. The modular body sections are: No 1 cab, radiator, power unit, turbocharger, electrical, No 2 cab. Cab modules are fully interchangeable.

Unusual for a main line locomotive the equipment modules are not accessible from the cabs, so the crews do not have any engine room duties to perform.

Construction went ahead at Doncaster Works and the initial order for 35 locomotives was extended to 50, with more expected to follow in due course. Unfortunately, the collapse of a great deal of Railfreight traffic during the 1984-85 miners' strike finally ended this hope, and the last locomotive, No 58050, entered service in February 1987. This last member of the class was fitted with an experimental system of separate excitation of the six traction motors, which it was hoped would enable the high tractive output of the power unit to produce some outstanding feats of load haulage, comparable to the Foster Yeoman '59s'. The equipment was removed in 1988.

The entire class is based at Toton depot and is used principally for haulage of coal in Nottinghamshire, South Yorkshire, and the East and West Midlands. Several teething troubles were experienced, particularly with sanding gear, AWS equipment, wheelslip control and some hydraulic equipment, and this caused some delay in getting the class to work. No 58002 was the first to become due for an 'F' exam at Doncaster, some five years after entering service. Although designed with foreign sales in mind, no export orders have been received. The class obviously has a long-term future, and cascading will follow the same pattern as Class 56 as newer locomotives are produced.

Below left:
No 58001 under construction at Doncaster during 1982, with the power unit and many of the auxiliaries already in place. Entry to the cabs is gained by doors placed immediately behind the cab itself, on the recessed running plate, leaving the cabs themselves with full-length flush sides. *S. R. Batty*

Below:
A Worcester-Bescot crew training run gets away from Worcester behind No 58004 on 15 May 1984. Early examples did not have the cabside 'Railfreight' transfers applied to most of the class. *P. J. Green*

Left:
Class 58 No 58011 passes Rookley Junction with a short Toton-Arpley Speedlink service on 6 September 1984. Unlike other locomotives, the Class 58s have the sealed-beam headlight fitted to the nearside of the cab front instead of in a central position. *S. Turner*

Right:
A well-filled Merry-go-Round train heads northwards along the Walsall-Rugeley line on 15 July 1986 hauled by No 58031. The pneumatic filter and pipework mounted on the bufferbeam looks vulnerable to accidental damage, and a better cab-front appearance would have resulted from recessing the headlight and marker lights and removing the multiple-control jumper cables to a lower level. *J. Whitehouse*

Below:
Because of their rarity on passenger workings, Class 58s are in demand for enthusiasts' specials. One such example was a special 11.59 Corby-Kettering working run in connection with the Ian Allan Network Day on 24 October 1987, seen near Glendon South Junction in the care of No 58002. *J. C. Baker*

8 General Motors 3,300hp Co-Co Class 59

Class: 59
Purpose: Foster Yeoman stone traffic
Introduced: 1986
Original numbers: 59001-59005
Total built: 5

Details of Nos 59001-59005
Engine: General Motors 645E3C
No of cylinders and cycle — 16 two-stroke
Max cont rated output — 3,300hp
Main alternator: EMD AR11 MLD D14A
Traction motors: General Motors D87
Performance: Max tractive effort — 122,000lb
Braking: Air
Brake force: 69 tonnes
Speed: 96km/h (60mph)
Tank capacities: Engine fuel — 919gal
Lubricating oil — 202gal
Water capacity — 212gal

The supply of crushed stone to the building and civil engineering industries of southern England has become big business over the last two decades, and Foster Yeoman has always been at the forefront of this highly competitive market. Its quarry at East Cranmore, in deepest Somerset, has an ultimate capacity for removing the Mendip Hills at the rate of nine million tons per year, and every ton has to be quarried, crushed, loaded and delivered at the lowest possible price. Competition is so intense that each ton of stone has to be delivered to the customer exactly when it is wanted and at the lowest possible price — cost per ton is the yardstick — and with such large amounts being quarried the company cannot afford to have this figure raised by even the smallest amount.

From the late 1970s Bristol Bath Road depot supplied six Class 56s to operate the Foster Yeoman traffic. Three trains were double headed daily, but the results were far from satisfactory. Unreliability caused up to 40% of the trains to be late, resulting in disgruntled customers and poor utilisation of the 289-strong fleet of wagons leased from Procor. It was the urgent need to improve this situation which prompted Foster Yeoman to look at the possibility of purchasing a small fleet of purpose-built locomotives which would be capable of single-heading trains of over 4,300 tons in weight. The operation of privately-owned locomotives on BR was a revolutionary idea, and some problems had to be faced. Whatever locomotives were eventually chosen would have to be serviced by BR staff to ensure that BR running standards were maintained, and some trade union negotiation would be needed to establish the arrangements for the operational and maintenance requirements of the new class. These problems could be foreseen and be expected to be settled in quite a straightforward manner, but in late 1984 the biggest problem was the most fundamental — who would build the locomotives? BREL could only offer the Class 58, which was not proving the success it had been hoped to be and had generated no interest from the export market, and in any case did not offer sufficient haulage capability. The British locomotive-building industry was not interested in a one-off order for a small number of machines. On to the stage then came the Electro-Motive Division of GM, an

Below:
Cabside cast numberplate of No 59002, also showing the cast metal plate giving details of weight, brake force, route availability and maximum speed. The positive and negative rake of the cab front can be seen.
Brian Morrison

Right:
No 59001 traverses the Westbury avoiding line with a Merehead-Theale working on 6 May 1986. *C. J. Tuffs*

Below right:
Soon after delivery, No 59003 prepares to leave Theale with a train of empty 102-ton hopper wagons on 20 February 1986.
B. J. Nicolle

W

YEOMAN

American organisation which offered to supply just four locomotives which it was confident would be able to handle all Foster Yeoman's traffic needs. Moreover, the use of the 'Super Series' wheelslip control system would provide very high levels of adhesion between wheel and rail and so avoid the need for any double heading, thus allowing four locomotives to do the job of six conventional ones, representing a saving of £2 million on initial purchase price.

The four locomotives were ordered early in 1985 and all were delivered to Southampton docks on 21 January 1986. Visually, the new Class 59 locos (as they are classified under the TOPS system) are unlike anything else operating on BR. Viewed from the front there is a passing resemblance to the Class 58 cab-front arrangement, but the roof is steeply pitched, a large plate is fitted below the buffers behind the coupling gear and brake connections and there are no visible jumper connections. A full-width body runs between the two cabs, and a side view clearly illustrates the high solebar level and also the positive and negative rake of the cab front. The Foster Yeoman livery of blue and silver grey is pleasingly applied, the overall silver grey being overlaid with blue bands along the length of the locomotive at cantrail and solebar levels. This latter band is carried around the cab fronts, across the yellow warning panels which reach up from buffer level only to handrail level, just below the cab windows.

During three years of operation the class has been very successful, fulfilling all Foster

Below:

Soon after naming, No 59004 *Yeoman Challenger* approaches Pewsey with the 07.20 Merehead-Theale on 21 July 1986. Conventional SSC equipment is fitted to the Class 59 locomotives for use at speeds of up to 4mph, and the associated sandboxes are clearly visible at the ends of each bogie.

T. Beaven

Yeoman's requirements. The 3,300hp 16-cylinder two-stroke engine is a piece of straightforward conventional diesel engineering which only needed a large modified silencer to enable BR's exhaust noise limits to be met. Inside the cabs, the driver's control desk is laid out similarly to the Class 58 arrangement. The 'Super Series' wheelslip control system lies at the heart of the tremendous haulage capacity of these locomotives. Generally speaking, it is installed engine horsepower and train braking ability which govern the speed of a train once it is rolling, but the starting ability is determined by the adhesion which can be obtained between the locomotive's driving wheels and the rails, often under adverse conditions. Adhesion generally varies between 10% and 40% on BR tracks, and along the route to be followed by the Foster Yeoman trains a value of 33-34% was estimated generally to be required for starting a loaded train under adverse conditions, with 23-24% being needed under normal running conditions. GM offered figures of 39% and 28% respectively for the Class 59s, and these figures meant that a single locomotive would easily be capable of hauling the planned trainloads of 4,300 tons. If double heading was ever needed (where the leading locomotive would 'condition' the track and provide even better adhesion for the second locomotive), then gross weights of 10,000 tons were estimated as being quite feasible for haulage out of Merehead. Wheelslip control is an established feature of diesel locomotive design, but the GM system has a particularly sophisticated feature which was developed during the 1970s and which took advantage of the latest advances in microprocessor and radar technology. Conventional systems rely on the detection of wheelslip by electrical coils and relays which respond to the change in traction motor voltage which occurs when slipping takes

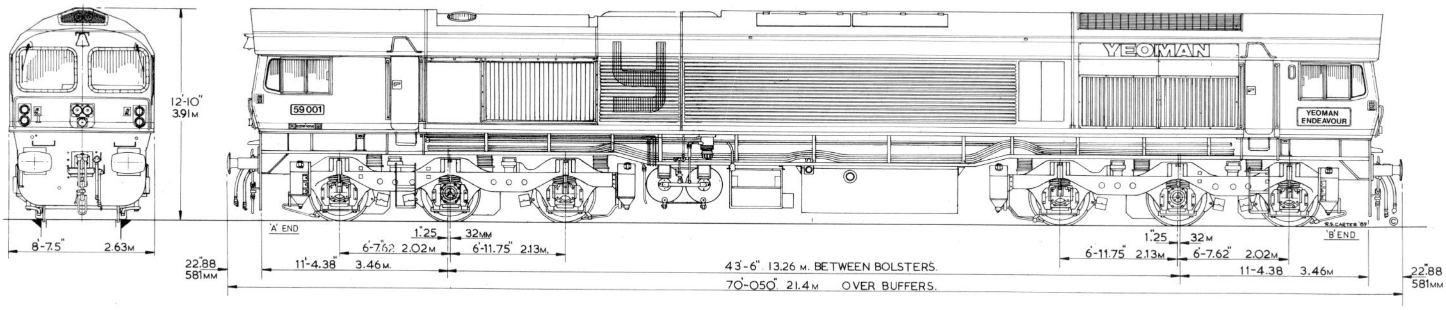

Above:
General arrangement of General Motors Class 59 3,300hp Co-Co. *R. S. Carter*

place — the power to that motor is briefly stopped, and then reapplied when slipping stops. Microprocessors improved the precision of this control, especially in the measurement of axle speed, but the critical factor was measurement of the actual locomotive speed. Maximum adhesion takes place within a narrow speed band where the wheel rotates faster relative to the rail at a rate 1%-10% greater than the forward locomotive speed. The GM system keeps this figure at 1.5% and uses a radar system precisely to measure the locomotive speed. The microprocessors control the power supply to the traction motors to keep the 'creep' speed at the 1.5% level. This corresponds to the point at which wheelslip is just about to start, and maximum grip is obtained at this point.

GM guaranteed an availability figure of 95% for the new fleet, and during the first year of operation a figure of 99.7% was achieved. To be fair to BR (and to avoid some burst blood vessels within BR's local management), this figure is based on the number of trains hauled over a five-day period, whereas BR's availability calculations are based on a full seven-day period including stoppages for maintenance. On this basis, the Class 59 guaranteed figure comes out at 67.8%, only slightly better than the 65% target for the BR fleet.

Loaded trains are worked from Merehead (where servicing work is carried out on a daily basis, in addition to a series of four-weekly examinations to smooth out the total workload of maintaining the fleet) to Acton, Botley, Brentford, Eastleigh, Wootton Bassett, Purfleet and Theale. The locomotives have been very successful and traffic has increased to the extent that a fifth machine was ordered in February 1988 for delivery in 1989. This extra traffic has been hauled by BR locomotives, but the cost penalties incurred were deemed to be too great. Double heading may become a reality after all, with trainloads of almost 9,000 tons being hauled to London for splitting into separate trains which would then be taken forward as required. Foster Yeoman's principal competitor in the stone business, ARC, borrowed No 59004 for a brief trial from Whatley quarry in December 1988 and the results were sufficiently impressive to make ARC think seriously about obtaining a similar fleet, instead of waiting for BR's new Class 60 to appear.

The four locomotives were named on 28 June 1986.

59001	*Yeoman Endeavour*	59003	*Yeoman Highlander*
59002	*Yeoman Enterprise*	59004	*Yeoman Challenger*

A fifth locomotive was scheduled for delivery in June 1989.

Below:
No 59002 *Yeoman Enterprise* runs along the Wylye Valley near Wishford with the 07.20 Merehead-Eastleigh on 29 October 1986. The massive silencer assembly can be seen in the locomotive's roof, just behind the leading cab. *G. F. Gillham*

9 Brush/Mirrlees/Procor 3,100hp Class 60

Class: 60
Purpose: Heavy freight traffic
Introduced: 1989
Original numbers: 60001-60100

Details of Nos 60001-60100
Engine: Mirrlees Blackstone 8MB875T
No of cylinders and cycle — 8 four-stroke
Max cont rated output — 3,100hp
Main alternator: Brush
Performance: Max tractive effort — 65,000lb
Braking: Air
Speed: 60mph

The first of BR's Class 60 freight locomotives are now being assembled, after four years of planning, policy-changing and generally rather frantic searching for a design of freight locomotive to take Railfreight well into the next century. Whatever sort of performer the new class turns out to be, the story behind the design is worth recalling.

In 1985, when the last-built Class 56 locomotives were hardly a year old and BREL Doncaster Works was still producing the Class 58s, BR gave the locomotive-building industry a glimpse of its requirements for the next 25 years. If the timescale was enough to make the manufacturers blink hard, the number of locomotives involved sent them rushing towards BR with open arms. The approximate requirements were expected to total 450 passenger and 850 freight diesel-electric units, split broadly as follows:

Class 48 A passenger/parcels locomotive with freight capability to take over similar duties to the existing Class 47 fleet. Small pilot batches would be produced by the late 1980s, with full production taking place during the 1990s.

Class 38 Pure freight locomotives, possibly to be built in two power ratings, with a pilot run of 20 being delivered 1986/87 and production starting from 1990. The actual rate of delivery would vary according to the demise of Classes 20 and 37, those two English Electric stalwarts which are well known for longevity and robustness!

Major improvements in diesel engine design have meant that 1,800-2,000hp can now be provided by a compact six-cylinder engine instead of the larger, heavier 12-cylinder units of the 1960s, and two hotly-favoured contenders for the Class 38 fleet were the Ruston RK270T and the Mirrlees MB275T, and a clutch of these well-tried units was installed into a series of modified Class 37 locomotives for trial purposes. Locomotives Nos 37901/904 were fitted with the Mirrlees engine and Nos 37905/06 with the GEC/Ruston unit, and all were rated at 1,800hp. Consideration was also given to engines from foreign manufacturers such as Krauss-Maffei and General Motors (GM).

The production schedule for the pilot batches slipped back over the succeeding 12 months and suddenly, in late 1986, the entire project (including Class 48) was shelved. Tough new demands from Central Government meant that those sectors of BR which did not benefit from Public Service Obligation (PSO) grants suddenly had to take a long, hard look at their abilities to give a vastly improved return on assets. The freight business felt the full blast of this chill wind and as its Speedlink traffic was nowhere near as lucrative as the bulk loads of coal, steel, stone and chemicals which BR had courted for many years past, the Class 38 project was quickly sidelined in favour of proposals for a new design of 3,000hp heavy-haul machine. By this time the Ruston and Mirrlees engines were entering service, and the GM locomotives leased by Foster Yeoman were providing levels of performance undreamt of in BR freight circles. Thus two manufacturers had both spent fortunes on gaining type approval for their engines, only to see the possible orders evaporate before their eyes, and Railfreight was frantically searching for a

Below:
The body for the first of the new Class 60 locomotives seen under construction at the Procor works, near Wakefield, on 12 January 1989. *L. A. Nixon, courtesy Procor/Brush*

brand-new design which could hopefully achieve a performance similar to the GM Class 59.

Specifications were quickly produced for the Class 60 locomotives, with a total of 100 expected to be built by 1992. BR wanted a high-performance locomotive built to a tight schedule and with strict warranty terms, and the possibility of short production runs deterred several foreign bidders from entering the fray. Two companies eventually emerged with the ability to deliver the new design — Brush and GEC. Brush would use either the Ruston or Mirrlees engines (both enlarged from six to eight cylinders, and capable of producing 2,750hp-plus), and GEC initially thought of using one of the above-mentioned units or perhaps a 12-cylinder unit as had been installed in the Class 58. The specification was tied down by this time to a preference for an in-line engine which gives much easier access for maintenance and was expected to be generally cheaper than a large vee-form unit, and further studies of the traffic and routes likely to be widely used by the Class 60 locomotives indicated that 2,750 installed horsepower would be adequate. Brush were acknowledged to be the builders of very good bogies and could produce a microprocessor-controlled separate-excitation system (SEPEX) to operate in conjunction with the well-tried and fuel-efficient engines. Metropolitan-Cammell offered to build a locomotive which would meet the BR requirements and would also be a 'state-of-the-art' product using a Ruston or Mirrlees engine and three-phase ac drive via traction motors mounted on Brush bogies, all of which would give performance levels equal to that of Class 59. Unfortunately, after further study of the specification, the company decided they would not be able to give their best to the design, and withdrew from the tendering process.

What, then, of the GEC offering at this stage? Despite having access to good engines (both of the newer Ruston and Mirrlees variety as well as the older 12- and 16-cylinder RK3 series), the company felt it had been out of the high-power diesel market for too long and did not want to enter the fray with its existing hardware. Neither did it want to use what it regarded as an obsolete ac/dc transmission (GEC's own three-phase system is still being developed), and the end result was a bid to build an anglicised, bought-in GM design, complete with V-16 two-stroke engine! Licensing agreements would allow GEC to build the electrical equipment and BREL to look after the mechanical items (apart from the engine), and BR would be given a well-proven and reliable design, albeit with a thirsty engine which had too many cylinders in the wrong configuration!

During early 1988 the approval was given to order up to 100 of the new locomotives from the Brush Co at Loughborough. The Ruston engine did not pass the type approval test but the eight-cylinder Mirrlees 8MB275RT did, and was therefore chosen as the unit to power the new class. Manufacture of the bodies was subcontracted to Procor of Wakefield, and assembly of all the major components for the first locomotive was well under way by January 1989, with an expected delivery date of June 1989 for No 60001.

General appearance is very much along the lines of the earlier Class 47 and 56 designs (not a hint of Class 58, apart from the driver's desk layout!), but a noticeable feature is the use of both positive and negative rake in the cab front design. From cab-front handrail level, the windows slope backwards to the roof and the lower panel also slopes backwards, but at a lesser angle, until it reaches bufferbeam level. A large plate, almost of snowplough appearance, is fitted below the buffers between the coupling gear and leading wheelset. Air intake grilles are fitted into the bodysides in a similar fashion to the Class 56. Two-tone grey Railfreight livery with appropriate sub-Sector and depot badging will be applied, and the cab fronts will have yellow paint from above the buffers up to the cab windows, which will have black-painted frames. A pair of highly-visible, chrome-plated air horns will be mounted in a recess on each cab roof.

Looking at the end product of the Class 60 saga and bearing in mind the turmoil surrounding its conception, one hopes that BR has not simply ended up with a simplified, more fuel-efficient variation on the Class 56 theme. It is expected that the 60mph Class 60 locomotives will cope with heavy loads of coal, steel, oil and stone far more effectively than the 3,250hp 80mph Class 56 or the 3,500hp combination of two Class 37s. The Railfreight Sector's recent takeover of the Freightliner business means that it must be able to operate a fast, long-distance distribution service involving considerable loads. Hence, Class 65 may appear in the not too-distant future (although at present it is still at the outline stage), with an installed horsepower of 4,000 provided by a Mirrlees ESL V-16 engine. Issues to be addressed will include the problems of total locomotive weight and cooler group capacity, just as happened some 25 years ago with the earlier prototype high-power locomotives, although today's more efficient cylinder design should ease the problems somewhat.

Meanwhile, the Class 38 design is still very much sidelined and Class 48 has disappeared altogether amidst InterCity's HST and electrification plans and the Provincial Sector's liking for the new generation of DMUs. Today's Classes 20/31/37 will eventually need replacement, and a medium-powered alternative will have to be found. The recent Class 41 proposal is for a locomotive of 2,200-2,500hp intended for use on coal traffic which still leaves room for a Type 3 replacement.

Below:
The first Class 60 to be completed is seen here on 30 June 1989 following the official handover. No 60001 has been named